Chemistry Research and Applications

Chemistry Research and Applications

A Closer Look at Carvacrol
Zak A. Cunningham (Editor)
2022. ISBN: 978-1-68507-627-6 (Softcover)

Fundamentals of Photocatalysis
Orva Auger (Editor)
2021. ISBN: 978-1-68507-374-9 (Softcover)
2021. ISBN: 978-1-68507-417-3 (eBook)

Polypropylene: Advances in Research and Applications
Théodore Marleau (Editor)
2021. ISBN: 978-1-68507-378-7 (Hardcover)
2021. ISBN: 978-1-68507-401-2 (eBook)

Boron: Advances in Research and Applications
Lynn Mcconnell (Editor)
2021. ISBN: 978-1-68507-231-5 (Softcover)
2021. ISBN: 978-1-68507-259-9 (eBook)

Applications of Layered Double Hydroxides
Rajib Lochan Goswamee, PhD (Editor)
Pinky Saikia, PhD (Editor)
2021. ISBN: 978-1-68507-355-8 (Hardcover)
2021. ISBN: 978-1-68507-381-7 (eBook)

Corrosion Inhibitors: An Overview
Raymond Wilkerson (Editor)
2021. ISBN: 978-1-68507-012-0 (Hardcover)
2021. ISBN: 978-1-68507-163-9 (eBook)

More information about this series can be found at
https://novapublishers.com/product-category/series/chemistry-research-and-applications/

Zak A. Cunningham
Editor

A Closer Look at Carvacrol

Copyright © 2022 by Nova Science Publishers, Inc.

All rights reserved. No part of this book may be reproduced, stored in a retrieval system or transmitted in any form or by any means: electronic, electrostatic, magnetic, tape, mechanical photocopying, recording or otherwise without the written permission of the Publisher.

We have partnered with Copyright Clearance Center to make it easy for you to obtain permissions to reuse content from this publication. Simply navigate to this publication's page on Nova's website and locate the "Get Permission" button below the title description. This button is linked directly to the title's permission page on copyright.com. Alternatively, you can visit copyright.com and search by title, ISBN, or ISSN.

For further questions about using the service on copyright.com, please contact:
Copyright Clearance Center
Phone: +1-(978) 750-8400   Fax: +1-(978) 750-4470   E-mail: info@copyright.com

NOTICE TO THE READER

The Publisher has taken reasonable care in the preparation of this book, but makes no expressed or implied warranty of any kind and assumes no responsibility for any errors or omissions. No liability is assumed for incidental or consequential damages in connection with or arising out of information contained in this book. The Publisher shall not be liable for any special, consequential, or exemplary damages resulting, in whole or in part, from the readers' use of, or reliance upon, this material. Any parts of this book based on government reports are so indicated and copyright is claimed for those parts to the extent applicable to compilations of such works.

Independent verification should be sought for any data, advice or recommendations contained in this book. In addition, no responsibility is assumed by the Publisher for any injury and/or damage to persons or property arising from any methods, products, instructions, ideas or otherwise contained in this publication.

This publication is designed to provide accurate and authoritative information with regard to the subject matter covered herein. It is sold with the clear understanding that the Publisher is not engaged in rendering legal or any other professional services. If legal or any other expert assistance is required, the services of a competent person should be sought. FROM A DECLARATION OF PARTICIPANTS JOINTLY ADOPTED BY A COMMITTEE OF THE AMERICAN BAR ASSOCIATION AND A COMMITTEE OF PUBLISHERS.

Additional color graphics may be available in the e-book version of this book.

Library of Congress Cataloging-in-Publication Data

ISBN: 978-1-68507-627-6

Published by Nova Science Publishers, Inc. † New York

Contents

Preface vii

Chapter 1   **Biological and Therapeutic Properties of Carvacrol** 1
A. Thalía Bernal-Mercado,
M. Melissa Gutiérrez-Pacheco,
Samaria L. Gutierrez-Pacheco,
Yessica Enciso Martínez, María González-Leyva,
Melvin R. Tapia-Rodríguez
and J. Fernando Ayala-Zavala

Chapter 2   **Carvacrol as an Additional Barrier for Control of Pathogens during the Thermal Processing of Meat** 53
Martin Valenzuela-Melendres,
María González-Leyva and Jorge I. López-Pino

Chapter 3   **Carvacrol as an Antibiofilm Agent in the Food Industry** 89
M. Melissa Gutiérrez-Pacheco,
Luis A. Ortega-Ramírez,
A. Thalía Bernal-Mercado,
Cristóbal J. González-Pérez,
Samaria L. Gutiérrez-Pacheco
and J. Fernando Ayala-Zavala

Chapter 4   **Carvacrol Emulsification: From Theory to Applications** 127
Alana G. de Souza and Derval S. Rosa

Chapter 5	**Carvacrol Encapsulation:**	
	Strategies, Preparation Methods, and Trends 149	
	Rafaela R. Ferreira and Derval S. Rosa	
Index	 169	

Preface

Carvacrol is a monoterpene phenol found in essential oils of aromatic plants such as thyme and oregano. It is used as a food additive and as a fragrance in cosmetic products. As explored in the five chapters of this book, carvacrol also has applications in health due to its antibacterial and antifungal effects. Chapter one includes an overview of the biological and therapeutic properties of carvacrol. Chapter two discusses carvacrol's potential as part of a barrier technology in the thermal processing of meat products to limit and control the growth of pathogenic microorganisms. Chapter three deals with the antibiofilm properties of carvacrol and its potential to control biofilm formation in meat, dairy, and fresh produce industries. Chapter four provides an in-depth review of the emulsification of essential oils that contain carvacrol, addressing various preparation methodologies and types of stabilizers. Lastly, chapter five details strategies, preparation methods, and trends in connection with carvacrol encapsulation.

Chapter 1 - Interest in natural substances has grown in the food, cosmetics, and pharmaceutical industries in recent years. Consumers demand safer products without additives related to health damage, and some patients seek effective and safer alternatives to some drugs that cause strong side effects. Essential oils and their main compounds are considered one of the most valuable plant products due to their wide range of bioactive activities of great interest in the scientific and clinical area, and this opens the door to the use of complementary natural treatments. Carvacrol is a monoterpene phenol, chemically known as 5-isopropyl-2-methyl phenol, abundant in the essential oil of some medicinal plants such as oregano, thyme, and savory. The US Food and Drug Administration recognizes this monoterpene as safe for use as a food additive; current investigations are being conducted to establish its potential therapeutic effects. Carvacrol also has useful properties for clinical applications such as antimicrobial, anti-inflammatory and anti-carcinogenic; however, human trials are still lacking. Carvacrol can scavenge free radicals, inhibit viral and microbial human pathogens, and it causes cytotoxic anti-apoptotic activity against lung, liver and breast cancer cells. In this perspective, this chapter analyses and discusses the antioxidant, antimicrobial, anti-inflammatory, antiproliferative, antimutagenic properties of carvacrol and its potential therapeutic uses.

Chapter 2 - The main guarantee against foodborne illness is exposure to heat to eliminate pathogens present in meat and meat products. Additional barriers, such as pH control, water activity, and the addition of antimicrobials such as carvacrol, are used to increase safety margins during heat treatments of food. Carvacrol is a monoterpene phenol produced by various aromatic plants, and it is used in low concentrations as a flavoring and seasoning ingredient for food and has significant antimicrobial activity against the principal pathogens. Carvacrol interacts with the meat matrix components, mainly fat, and has a significant impact on the heat treatments of food. This compound affects the behavior of pathogens during the thermal process, making them more susceptible to the effect of temperature. However, since the direct use of carvacrol affects the product's sensory properties, it is necessary to use technologies such as atomization microencapsulation for its incorporation. The use of carvacrol as part of barrier technology in the thermal processing of meat products is a very promising strategy to limit and control the growth of pathogenic microorganisms.

Chapter 3 - The ability of microorganisms to form biofilms is a growing concern in the food industry. Food processing areas include an environment that combines moisture and nutrients and becomes ideal for biofilm development. Biofilms are communities of microorganisms embedded in a self-produced matrix of extracellular polymeric substances, which protect against environmental stresses. The major problems associated with biofilms are their persistence and resistance to cleaning and sanitizing procedures, which represent high economic losses and public health implications due to increased food spoilage and outbreaks. Although synthetic antimicrobials are approved in many countries, the trend has been the use of natural disinfectants, which fulfill the needs of today's consumers looking for safer, effective, and acceptable alternatives. Plant extracts such as essential oils represent a valuable source of biologically active molecules possessing antimicrobial properties. Specifically, carvacrol, the primary terpene compound of oregano essential oil, has shown high antimicrobial activity and inhibits biofilms of many food pathogens such as Escherichia coli, Salmonella Typhimurium, Listeria monocytogenes, Campylobacter jejuni, among others. Therefore, this chapter discusses the antibiofilm properties of carvacrol and their potential to control biofilm formation in meat, dairy, and fresh produce industries.

Chapter 4 - Carvacrol [2-methyl-5-(1-methylethyl)-phenol] is a monoterpene found in essential oils of aromatic plants with numerous active properties: antibacterial, antioxidant, antifungal, and anti-cancer. However, essential oils have high volatility, hydrophobicity, intense aroma, photosensitivity, low stability, high susceptibility to oxidation, and poor water solubility, limiting their application. Emulsification is considered a simple way to overcome these limitations. Emulsions are formed by homogenization of two immiscible liquid phases in which one of them is dispersed in the other in the form of tiny droplets. Emulsion systems are usually applied in hydrophilic-lipophilic mixtures, such as essential oil and water, to improve the oil solubility and facilitate application. However, an emulsification agent is needed to break the drops of the oily phase into smaller and more stable particles since oil and water do not coexist harmoniously in the same system. Oil molecules have instantaneous dipoles related to the molecule's chemical structure and functional groups, which reorient themselves all the time, generating constant forces of attraction or dispersion. One way to generate emulsions and overcome the energy barrier of the system associated with interfacial tension is through shear energy, seeking to balance the system's free energy. However, the stability is low, and there is a reversibility trend, resulting in coalescence in short periods. Additives that reduce the interfacial tension and the energy barrier, such as surfactants and solid particles, are an alternative to overcome this energy barrier and make the emulsions thermodynamically favored. The additives are adsorbed to oil/water interfaces, forming thick interfacial layers to stabilize the emulsions via electrostatic or steric mechanisms, resulting in droplets with micro or nanosizes with high stability. This chapter describes the theoretical aspects of micro and nanoemulsions formation, considering fluid properties, physical and chemical characteristics, and thermodynamic equilibrium. In addition, the authors do an in-depth review of the emulsification of essential oils that contain carvacrol, addressing various preparation methodologies and types of stabilizers. The solid particles, which make up the well-known Pickering emulsions, will be highlighted among the stabilizers. The state of the art of new technologies and applications is widely presented and discussed, such as for food, biomedical, delivery carrier systems, among others.

Chapter 5 - Essential oils (EO) have attracted significant attention as additives in chemical products (food) due to their olfactory, physical-chemical, and biological characteristics. The antimicrobial and antioxidant effects of oregano essential oil (OEO) are mainly attributed to the presence of carvacrol (phenolic monoterpene), which constitutes about 78-82% of the

Origanum vulgare plant essential oil. However, essential oils (EO) are unstable compounds and susceptible to degradation when exposed to environmental stresses such as oxygen, temperature, and light. Thus, alternatives are sought to improve their application; in this sense, encapsulation is a method that offers a viable strategy to stabilize and prolong the EO release, allowing their application as an additive to food products. Encapsulation is defined as a method in which minuscule particles or droplets are surrounded by a cladding wall or embedded in a matrix. The matrix wall isolates the active compound as a functional barrier from the surrounding environment to prevent chemical and physical reactions and prolong their stability and bioavailability. Colloidal systems promote encapsulation, and several factors influence the effectiveness of the process, such as the type of technique employed, emulsifier type and concentration, and wall material, reflecting on the encapsulation efficiency, particle size, and physical stability of encapsulated EOs. In addition, the associated benefits after encapsulation must be considered, that is, bioavailability, controlled release, and protection of the EO against environmental stresses. Thus, this chapter summarizes the recognized benefits, functional properties of various preparation and characterization methods, in which innovative manufacturing strategies and their mechanisms are demonstrated.

Chapter 1

Biological and Therapeutic Properties of Carvacrol

A. Thalía Bernal-Mercado[2],
M. Melissa Gutiérrez-Pacheco[1],
Samaria L. Gutierrez-Pacheco[1],
Yessica Enciso Martínez[1], María González-Leyva[1],
Melvin R. Tapia-Rodríguez[1]
and J. Fernando Ayala-Zavala[1,*]

[1]Coordinacion de Tecnologia de Alimentos de Origen Vegetal, Centro de Investigacion en Alimentacion y Desarrollo, Hermosillo, Sonora, Mexico
[2]Departamento de Investigacion y Posgrado en Alimentos, Universidad de Sonora, Hermosillo, Sonora, Mexico

Abstract

Interest in natural substances has grown in the food, cosmetics, and pharmaceutical industries in recent years. Consumers demand safer products without additives related to health damage, and some patients seek effective and safer alternatives to some drugs that cause strong side effects. Essential oils and their main compounds are considered one of the most valuable plant products due to their wide range of bioactive activities of great interest in the scientific and clinical area, and this opens

---

[*] Corresponding Author E-mail: jayala@ciad.mx.

In: A Closer Look at Carvacrol
Editor: Zak A. Cunningham
ISBN: 978-1-68507-627-6
© 2022 Nova Science Publishers, Inc.

the door to the use of complementary natural treatments. Carvacrol is a monoterpene phenol, chemically known as 5-isopropyl-2-methyl phenol, abundant in the essential oil of some medicinal plants such as oregano, thyme, and savory. The US Food and Drug Administration recognizes this monoterpene as safe for use as a food additive; current investigations are being conducted to establish its potential therapeutic effects. Carvacrol also has useful properties for clinical applications such as antimicrobial, anti-inflammatory and anti-carcinogenic; however, human trials are still lacking. Carvacrol can scavenge free radicals, inhibit viral and microbial human pathogens, and it causes cytotoxic anti-apoptotic activity against lung, liver and breast cancer cells. In this perspective, this chapter analyses and discusses the antioxidant, antimicrobial, anti-inflammatory, antiproliferative, antimutagenic properties of carvacrol and its potential therapeutic uses.

Keywords: natural compound, antimicrobial, antioxidant

Introduction

Plants are a rich source of several biologically active compounds with potential use in the health, food, cosmetics, and pharmaceutics industries (Hassan et al., 2018). Since ancient times, humans have used medicinal plants to alleviate and treat some diseases, and plant-derived molecules and their bioactive properties have encouraged the synthesis of drugs for decades (Potterat & Hamburger, 2008). Although traditional medicine is the most common practice, nowadays, the market of herbal medicine has grown worldwide as a form of prevention or alternative therapy, mostly in combination with conventional drugs (Enioutina et al., 2017). Consumers demand more natural and safer products because some traditional medicines may cause strong side effects and multidrug resistance (J. Zhang, Onakpoya, Posadzki, & Eddouks, 2015). With this in mind, numerous studies have driven the search for bioactive compounds with promising therapy functions.

Essential oils (EOs) are secondary metabolites extracted from medicinal and aromatic plants, composed of volatile terpenes and hydrocarbons (Baptista-Silva, Borges, Ramos, Pintado, & Sarmento, 2020). EOs and their main compounds are of great interest in the research and clinical area owing to their valuable properties (Ahmadi, Fazilati, Nazem, & Mousavi, 2021; Ragno et al., 2020; Simirgiotis et al., 2020). Carvacrol is among EOs constituents with several biological activities claimed by scientific studies

(Marinelli, Di Stefano, & Cacciatore, 2018; Sharifi-Rad et al., 2018). Carvacrol is an oxygenated monoterpene constituted by a phenolic ring with methyl and isopropyl substitutions. It is present in the EOs of some medicinal plants such as oregano (*Origanum vulgare, Lippia graveolens)*, thyme (*Thymus vulgaris*), and savory (*Satureja hortensis*), among others (Can Baser, 2008; Côté, Pichette, St-Gelais, & Legault, 2021; Krisilia, Deli, Koutsaviti, & Tzakou, 2021; Marinelli et al., 2018; Rota, Herrera, Martínez, Sotomayor, & Jordán, 2008; Zgheib et al., 2019). Carvacrol ($C_{10}H_{14}O$) with a molecular weight of 150.22 g/mol is chemically named 2-methyl-5-(1-methyl ethyl)-phenol by the International Union of Pure and Applied Chemistry. This terpenoid is a colorless to pale yellow liquid, insoluble in water but soluble in ethanol, acetone, and diethyl ether. It has lipophilic properties (Log P=3.40) and a density of 0.976 g/mL at 20°C (Nostro & Papalia, 2012; Sharifi-Rad et al., 2018) (Figure 1).

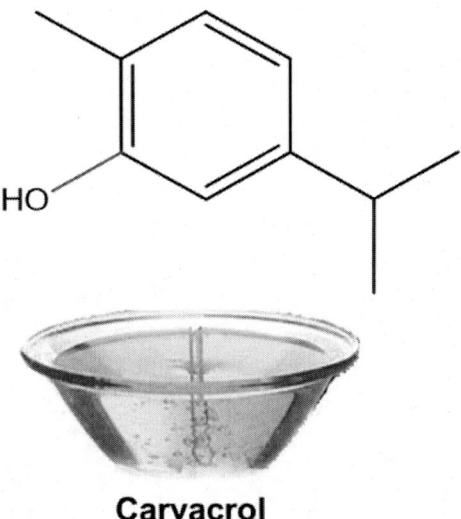

Carvacrol

Figure 1. Carvacrol chemical structure and physical aspect.

Carvacrol is generally recognized as safe by the Federal Drug Administration as a food additive in numerous products such as baked goods, meat, gelatin, alcoholic and non-alcoholic beverages (De Vincenzi, Stammati, De Vincenzi, & Silano, 2004). Also, it is included in the list of chemical flavorings of the Council of Europe and used in washing and cleaning products, biocides, fragrances, cosmetics, and personal care products (Suntres,

Coccimiglio, & Alipour, 2015). According to the European Chemicals Agency, the acute toxicity of carvacrol in rats is LD_{50} = 810 mg/kg of body weight (bw) when administrated orally. The carvacrol LD_{50} in mice was 80.00, 73.30, and 680 mg/kg bw when administrated intravenously, intraperitoneally, and subcutaneously, respectively (Andersen, 2006). Another study indicated that one-month treatment with carvacrol at a dose of 1mg/kg/day and 2 mg/kg/day showed clinical safety and tolerability for this agent in healthy persons (Ghorani, Alavinezhad, Rajabi, Mohammadpour, & Boskabady, 2021).

Carvacrol is widely used in food, cosmetic, and fragrance products; however, its use in clinical applications is still limited. Current investigations are searching to establish the potential therapeutic effects of carvacrol to prevent and alleviate diseases (Suntres et al., 2015). Carvacrol possesses many biological and pharmacological properties mainly attributed to its structure and characteristics: a free hydroxyl group, hydrophobicity, and a phenol moiety (Sharifi-Rad et al., 2018). In addition, studies have reported that carvacrol has antimicrobial activity against *Escherichia coli, Pseudomonas aeruginosa, P. fluoresencens, Staphylococcus aureus, Salmonella typhimurium, Listeria monocytogenes, Pectobacterium carotovorum, Bacillus subtilis, Saccharomyces cerevisiae, Botrytis cinerea, Candida albicans,* and others (Ben Arfa, Combes, Preziosi-Belloy, Gontard, & Chalier, 2006; Engel, Heckler, Tondo, Daroit, & da Silva Malheiros, 2017; Gutierrez-Pacheco et al., 2018; Swetha, Vikraman, Nithya, Hari Prasath, & Pandian, 2020; Tapia-Rodriguez et al., 2019). Also, it can inhibit some bacterial virulence factors such as motility, biofilm formation, and *quorum sensing*, which may help treat microbial infections (Gutierrez-Pacheco et al., 2018; Tapia-Rodriguez et al., 2019). Furthermore, carvacrol has demonstrated great antioxidant potential with hepatoprotective effect, DNA protection, and reduction of oxidative stress damage, which could be used to prevent some diseases (Khalaf et al., 2021; Samarghandian, Farkhondeh, Samini, & Borji, 2016; Ündeğer, Başaran, Degen, & Başaran, 2009).

Carvacrol has a significant effect in the anti-inflammatory process due to its ability to suppress the expression of cyclooxygenase and inhibit the production and actions of nitric oxide (Ezz-Eldin, Aboseif, & Khalaf, 2020; Hotta et al., 2010; Mahmoodi et al., 2019). Many *in vitro* and few *in vivo* studies have reported that carvacrol exhibits strong genotoxic, cytotoxic and proapoptotic activity against lung, breast, liver, colon, and prostate cancer cells (Heidarian & Keloushadi, 2019; F. Khan, Khan, Farooqui, & Ansari, 2017; I. Khan, Bahuguna, Kumar, Bajpai, & Kang, 2018; Mari et al., 2020;

Sharifi-Rad et al., 2018; Shinde, Agraval, Srivastav, Yadav, & Kumar, 2020). However, carvacrol application in the pharmaceutical area could be compromised considering its volatile nature, low aqueous solubility, poor bioavailability, hydrophobic and irritant characteristics (Shinde et al., 2020). Recent studies have developed strategies such as compound combination and drug-controlled release systems (Niaz, Imran, & Mackie, 2021; Shinde et al., 2020; Trindade et al., 2019). In this perspective, this chapter aimed to review the existing literature concerning the antioxidant, antimicrobial, anti-inflammatory, antiproliferative and antimutagenic properties of carvacrol and its potential use as treatments.

Extraction and Isolation of Carvacrol

Carvacrol is naturally found in the EO of some aromatic and medicinal plants, and it is related to its odor, flavor, and bioactive properties. This compound is one of the main constituents in several plants like oregano, thyme, savory, and others resumed in Table 1 (Can Baser, 2008; Côté et al., 2021; De Vincenzi et al., 2004; Marinelli et al., 2018; Rodriguez-Garcia et al., 2016; Zgheib et al., 2019). The composition of plant EOs and hence carvacrol content may vary according to different plant sources, environmental and climatic factors, plant genetic, season collection period, and extraction technique (Morshedloo, Salami, Nazeri, Maggi, & Craker, 2018; Pourhosseini, Ahadi, Aliahmadi, & Mirjalili, 2020).

Various techniques have been developed to extract plant EOs, such as distillation with steam pressure, hydrodistillation, and non-polar solvent extraction, which are the highly conventional procedures for this purpose (Tongnuanchan & Benjakul, 2014). It is essential to consider the plant organ (commonly aerial organs), the solvents, time, temperature, and processes used to obtain a high-quality EO since the extraction method can modify the chemical, quantity, and bioactive properties of EOs (Aziz et al., 2018). Several studies have reported the successful extraction of carvacrol-rich EOs employing conventional methods. Krisilia et al. (2021) obtained an EO from dried aerial parts of *S. thymbra* rich in carvacrol (65.2%) using hydrodistillation for three hours with a modified Clevenger-type apparatus. Also, Soto-Armenta et al. (2017) obtained an EO from *L. graveolens* leaves with carvacrol as the major component (30.89 – 36.24%) applying steam distillation. On the other hand, González-Trujano et al. (2017) extracted

Mexican oregano EO with a carvacrol content of 6.75% using hexane, ethyl acetate, and methanol as a solvent extraction technique.

Table 1. Main sources of carvacrol

Essential oil	Plant species	Carvacrol content (%)	Extraction technique	References
Oregano	Origanum vulgare	68	Supercritical CO_2	(Rodrigues et al., 2004)
	O. majorana	56.40 - 86.47	HD	(Bagci, Kan, Dogu, & Çelik, 2017)
	O. ehnrebergii	48.1 - 88.6	HD	(Zgheib et al., 2019)
	O. syriacum	82.8 - 86.8	Ultrasonic	(Novak, Lukas, & Franz, 2010)
	O. onites	86.9	HD	(Baydar, Sağdiç, Özkan, & Karadoğan, 2004)
	Lippia graveolens	30.89 - 36.24	SD	(Soto-Armenta et al., 2017)
Thyme	Thymus vulgaris	12	HD	(Gavaric, Mozina, Kladar, & Bozin, 2015)
	T. hyemalis	40.1	SD	(Rota et al., 2008)
	T. zygis	3.5	SD	(Khadir et al., 2016)
Savory	Satureja hortensis	67	HD	(Mihajilov-Krstev et al., 2009)
	S. thymbra	42.7	SD	(Özkan et al., 2017)
Bee balm	Monarda didyma	49.03	SD	(Côté et al., 2021)
	M. fistulosa	10	Methanol extraction	(Shanaida, Jasicka-Misiak, Bialon, Korablova, & Wieczorek, 2021)
Thymbra	T. spiccata	75.5	HD	(Baydar et al., 2004)
	T. calostachya	73.6 – 78.4	HD	(Krisilia et al., 2021)
	T. capitata	50.4 – 78.1	HD	(Krisilia et al., 2021)
Zataria	Zataria multiflora	57	HD	(Kavoosi & Rabiei, 2015)

*HD = hydrodistillation, SD = steam distillation.

Although conventional methods are the most straightforward and most frequently used procedures for EO extraction, literature research has reported some disadvantages, including long extraction time, low efficiency, and use of large amounts of toxic solvents (Knez Hrnčič et al., 2020; Sahraoui, Hazzit, & Boutekedjiret, 2017). For this reason, emerging technologies with high efficiency, shorter extraction time, economical and environmentally friendly characteristics have been used to obtain EOs with good bioactive compounds content (Hashemi, Khaneghah, & Akbarirad, 2016; Tongnuanchan & Benjakul, 2014).

Table 2. Conventional and alternative essential oil extraction techniques

Methods	Characteristics	Disadvantages	Reference
Hydrodistillation	Use steam at high temperatures to recover volatile aromatic compounds. This method involves the immersion of the plant material directly in boiling water.	Long procedure extraction times and potential hydrolysis of heat-sensitive compounds.	(Tongnuanchan & Benjakul, 2014)
Stem distillation	Use steam of heating water at high temperatures to recover volatile aromatic compounds from plants (placed in another container).	Lengthy procedure extraction times and potential hydrolysis of heat-sensitive compounds.	(Aziz et al., 2018)
Solvent extraction	Use solvents such as methanol, ethanol, acetone at mildly heating to extract essential oil from plant material. This method is used for sensitive heat plant materials.	Toxic solvent residues.	(González-Trujano et al., 2017)
Supercritical fluid extraction	Use mainly CO_2 under high-pressure conditions and heating to extract volatile compounds from plant materials. This method is fast, uses moderate temperatures and free-toxic solvents.	Require specific equips that apply high pressure (1,000 - 5,000 psi) required maintaining CO_2 solvent in supercritical conditions.	(Busatta et al., 2017)
Solvent-free microwave extraction	Heats plant materials using a microwave process and then uses a dry distillation. This technique is performed at atmospheric conditions without adding any solvent or water.	The extraction is from fresh plant materials, which are not easy to preserve, or prior moistened dried materials which make a complex process.	(Z. Wang et al., 2006)
Ultrasound extraction	Comprises the generation of cavitation bubbles that explode at the surface of the plant cell tissue to destroy the oil glands improving the mass transfer between the cell and the solvent to release the essential oil.	Temperature, solvent levels, and formation of bubbles need to be monitored.	(Hashemi et al., 2018)

These alternative methods include supercritical fluids extraction, solvent-free microwave extraction, microwave and ultrasound-assisted extraction, and other promising techniques (Aziz et al., 2018; Busatta et al., 2017; Sahraoui et al., 2017). Table 2 shows the main characteristics of conventional and alternative extraction techniques for obtaining EOs. Some studies have reported the higher efficacy of emerging methods to obtain EOs compared to conventional techniques. For example, supercritical CO_2 extraction was more effective than hydrodistillation regarding the higher yield of oregano EO and higher carvacrol content (Busatta et al., 2017).

Similarly, the solvent-free microwave extraction showed higher carvacrol content, lower extraction time, and higher saving energy than hydrodistillation to obtain *T. vulgaris* EO (Khalili, Mazloomifar, Larijani, Tehrani, & Azar, 2018). Tavakolpour et al. (2017) reported the influence of different extraction techniques on the air-dried aerial part EO of *Thymus daenersis*. Hydrodistillation showed the most extensive extraction time and lowest EO yield, while the ultrasound-assisted ohmic extraction showed the shortest time extraction and the higher yield. In addition, the authors observed a different chemical profile of EOs among techniques extractions, but only a slightly higher carvacrol content could be appreciated with ohmic techniques.

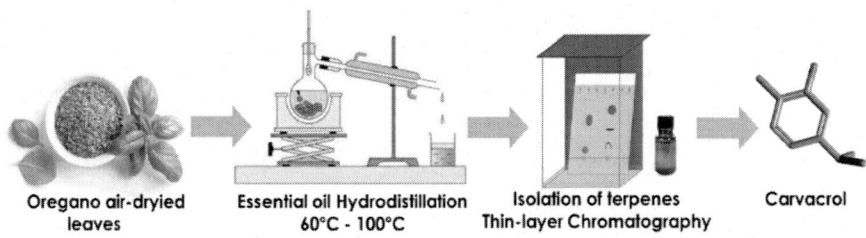

Figure 2. Carvacrol isolation process from oregano plant material.

Carvacrol can be isolated from plant EO by analytical procedures. For this, other EO compounds need to be removed by using solvents and chromatographic techniques such as column or thin-layer chromatography. An example of this protocol is using NaOH solutions dissolved in pentane, then silica gel plates with hexane and ethyl acetate as mobile phase to obtain carvacrol fraction (Figure 2) (Galehassadi, Rezaii, Najavand, Mahkam, & Mohammadzadeh, 2014).

Finally, the quantification of carvacrol can be analyzed by gas chromatography-mass spectroscopic, high-performance liquid chromatography (HPLC), and nuclear magnetic resonance (NMR) by the identification of its phenolic ring and hydroxyl group (Mutlu-Ingok, Catalkaya, Capanoglu, & Karbancioglu-Guler, 2021). However, the isolation and purification of EO compounds require long time and special procedures for separation, detection, and identification (Knez Hrnčič et al., 2020).

Biological Activities of Carvacrol

Antimicrobial Effects

The antimicrobial activity of EOs and their main constituents have been reported since ancient years. Our ancestors used aromatic plants in traditional medicine for their preservative and therapeutic properties in treating different diseases due to their biological activities. Both EOs and their volatile constituents are crucial compounds for biomedical or pharmaceutic purposes due to their antimicrobial (i.e., bactericidal and fungicidal) and medicinal (i.e., analgesic, sedative, anti-inflammatory, spasmolytic, and locally anesthetic remedies) properties (Raut & Karuppayil, 2014). In this sense, the use of plant-based therapeutics with improved antimicrobial activity and less toxicity is increasingly being accepted as alternatives to conventional antibiotic therapy (Buchbauer & Bohusch, 2016).

EOs components could represent an exciting approach to control the spread of pathogenic microorganisms. In this context, carvacrol has emerged for its broad spectrum of antimicrobial activity against Gram-positive and Gram-negative bacteria and fungi and has been investigated by many researchers worldwide (Gutierrez-Pacheco et al., 2018; Tapia-Rodriguez et al., 2019). The use of carvacrol as an antimicrobial compound has been exploited in different areas such as food hygiene, cosmetics, and medicine (Raut & Karuppayil, 2014). Furthermore, the broad spectrum of antimicrobial activity could allow its use to treat different types of infections due to its ability to inhibit oral, gastrointestinal, and urinary tract pathogens, among others.

Oral Microorganisms
Bacteria can be found on all oral tissues and constitutes part of the microbiota, which helps maintain the oral cavity's homeostasis and health. The disturbance of this microbiota composition causes a series of oral infectious diseases,

including dental caries, apical periodontitis, periodontal diseases, among others (Mosaddad et al., 2019). Dental caries and periodontal disease are associated with diverse oral pathogens and are commonly attributed to their capacity to form biofilms on teeth surfaces and supportive tissues (Karygianni et al., 2016) (Figure 3). For example, *Streptococcus mutans* is one of the main microorganisms associated with dental caries; it forms a biofilm and secretes acids that demineralize the tooth (Mosaddad et al., 2019). On the other hand, three oral anaerobic bacteria, including *Porphyromonas gingivalis*, *Treponema denticola*, and *Tannerella forsythia* have conventionally been regarded as causative factors of periodontitis (G. P. Wang, 2015).

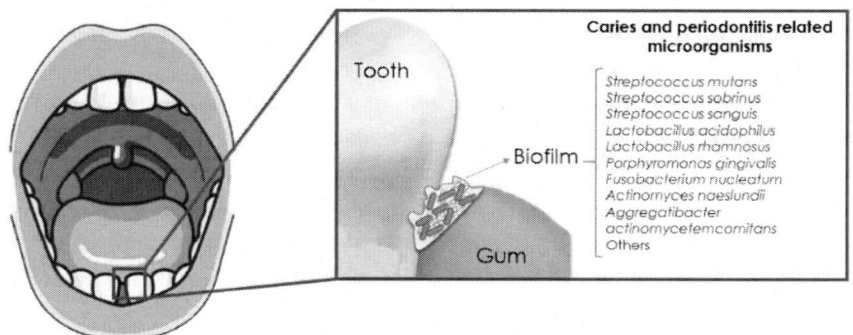

Figure 3. Dental caries and periodontitis disease related to microorganisms.

The effect of carvacrol to inhibit the growth of oral pathogens has been widely reported. S. T. Khan et al. (2017) reported that carvacrol reduced the viability, metabolic activity, and biofilm formation of dental caries caused by *S. mutans* at a concentration of 100 µg/mL. Similarly, T.-H. Wang et al. (2016) reported that carvacrol affects the viability of *S. mutans* and *Aggregatibacter actinomycetemcomitans*, other causal agents of dental caries and periodontitis, at concentrations of 200 and 600 µg/mL. Also, methicillin-resistant *S. aureus* (MRSA) (bacteria that colonize 77.8% of oral cancer patients after surgery) was inhibited at a 600 µg/mL concentration. Ciandrini et al. (2014) reported minimal inhibitory concentrations (MIC) of 0.25% (v/v) against the oral pathogens *P. gingivalis* and *Fusobacterium nucleatum*. Also, their capacity to form biofilms was affected at 0.5% for *S. mutans* and 1.0% for *P. gingivalis* and *F. nucleatum*.

Carvacrol also showed a fungicidal activity against *C. albicans* at 0.1% (Chami, Bennis, Chami, Aboussekhra, & Remmal, 2005). Carvacrol also significantly inhibited the growth of the cariogenic strains *S. mutans*, *S. mitis*, *S. salivarius*, *S. sanguinis*, and *C. albicans* at 50 mg/mL (Botelho et al., 2007).

On the other hand, Obaidat, Bader, Al-Rajab, ABU SHEIKHA, and Obaidat (2011) developed mucoadhesive patches containing tetracycline hydrochloride and carvacrol as a novel oral drug delivery system for the treatment of mouth bacterial infections and candidiasis. These authors reported that discs containing carvacrol alone showed excellent activity against *C. albicans* in amounts ≥40 µg/disc, whereas in combination with tetracycline were effective against all tested microorganisms such as *C. albicans*, *S. aureus*, *B. cereus*, *B. bronchis*, *E. coli*, and *P. aeruginosa*. Based on previous knowledge about its antimicrobial properties, carvacrol and other EOs constituents are used commercially against oral pathogens in mouthwashes and other treatments.

Gastrointestinal Microorganisms

Gastrointestinal infections continue to cause illness and death and contribute to economic loss in most parts of the world (Ternhag, Törner, Svensson, Ekdahl, & Giesecke, 2008). It has been estimated that they are the second most common infectious diseases after respiratory tract infections and a significant cause of morbidity and mortality among infants and children (WHO, 2017). Gastrointestinal infections can be caused by many microorganisms such as viruses, protozoa, fungi, and bacteria, the most common causal agents. Rotavirus, *Giardia lamblia, Salmonella* species, *Shigella* species, *Clostridium difficile, Campylobacter jejuni,* and *E. coli* are the most common etiological agents (Negrut et al., 2020). Gastrointestinal infections have been mainly associated with consuming food or water contaminated with feces and exposure to pathogens present in the hospital environment (Negrut et al., 2020).

Carvacrol can be used as a treatment of gastrointestinal diseases because different studies have shown its antimicrobial activity. Table 3 showed different studies where carvacrol has been effective against gastrointestinal pathogens using model systems.

Although few studies about its use in treating these infections, carvacrol has also been tested on different foods and food contact surfaces and has shown potent antimicrobial activity (Table 4).

Considering that the consumption of contaminated food causes a large part of gastrointestinal diseases, the preventive use of carvacrol becomes relevant to prevent food contamination with gastrointestinal pathogens and its direct use as a therapeutic agent.

Table 3. Antimicrobial activity of carvacrol against gastrointestinal pathogens in animal and cell systems

Microorganism	Concentration	Model	Results	References
Clostridium difficile	0.05 and 0.1% in combination with antibiotics and clindamycin	Mice	Carvacrol supplementation significantly reduced the incidence of diarrhea and positively altered the microbiome composition	(Mooyottu et al., 2017)
Campylobacter jejuni	0.5 mg/mL	Mice	Reduced pathogen loads in intestines and disease symptoms	(Mousavi et al., 2020)
	0.001 and 0.002%	Human intestinal epithelial cells (Caco-2)	Reduced attachment and invasion of *C. jejuni* to intestinal epithelial cells and the transcription of genes critical for infection in humans	(Upadhyay et al., 2017)
Escherichia coli O157:H7	0.5 mg/mL	Rumen system	Reduced *E. coli* levels from 10^3 and 10^6 CFU/mL to undetectable levels after 24 h incubation	(Rivas et al., 2010)
Rotavirus	391.5 µg/mL	Mardin-Darby bovine kidney (MDBK) cells	Carvacrol inhibited rotavirus in infected MDBK cells	(Pilau et al., 2011)
Escherichia coli STEC	0.0156%	Human ileocecal carcinoma cell line HCT-8	Carvacrol interfere with the bacterial colonization process and reduce adherence to HCT-8 cells	(Stratakos et al., 2018)

Table 4. Antimicrobial activity of carvacrol against gastrointestinal pathogens in food and food contact surfaces

Microorganism	Food/Surface	Concentration and application form	Results	References
Salmonella spp.	Stainless steel	0.33 mg/mL	Reduction of *Salmonella* growth and attachment	(Engel et al., 2017)
Salmonella Typhimurium biofilms	Polystyrene and stainless steel	0.05-0.10%	7-log CFU biofilm mass reduction to a nondetectable level at 1 and 4 h of exposure time for stainless steel and polystyrene surfaces, respectively	(Soni et al., 2013)
Salmonella Enteritidis *and Escherichia coli* O157:H7	Mung bean and alfalfa seeds	0.4% (nanoemulsion)	Treatment for 60 min inactivated *S.* Enteritidis and *E. coli* O157:H7. Pathogens were not detected after seed sprouting	(Landry, Chang, McClements, & McLandsborough, 2014)
Escherichia coli O157:H7	Apple, mango, orange, and tomato juices	1.33 mM	Carvacrol in combination with heat at 54°C reduce in 75% the time needed to inactivate 5 $\log_{10}$ cycles of *E. coli* O157:H7 in juices	(Ait-Ouazzou, Espina, García-Gonzalo, & Pagán, 2013)
Campylobacter jejuni	Chicken skin	2%	Carvacrol suspension wash reduced *C. jejuni* counts by ~2.4 to 4 $\log_{10}$ CFU/sample	(Shrestha et al., 2019)
Listeria monocytogenes	Raw catfish fillets	2%	Carvacrol solution reduced total microbial load from catfish fillets by approximately 5 log CFU/g	(Desai, Soni, Nannapaneni, Schilling, & Silva, 2012)
Pseudomonas aeruginosa	Polypropylene	0.06%	*P. aeruginosa* biofilm was reduced by 4.79 log CFU/cm^2	(Ashrafudoulla, Rahaman Mizan, Park, & Ha, 2021)

Respiratory Microorganisms

Globally, respiratory infections are the leading cause of morbidity and mortality from infectious diseases worldwide. More than one billion people are affected by acute and chronic respiratory diseases. These infections mainly affect infants, children, and older people (Murdoch & Howie, 2018) and are typically caused by viruses, bacteria, or mixed viral–bacterial infections, which can be contagious and spread rapidly through respiratory droplets. The main causal agents of respiratory infections are rhinovirus, coronavirus, parainfluenza, adenovirus, and bacteria such as *Haemophilus influenzae*, *Streptococcus pyogenes*, *Streptococcus pneumoniae*, *Moraxella catarrhalis* (N. Zhang et al., 2020).

Some authors have evidenced the potential of carvacrol to inhibit different respiratory pathogens. This effect is the case of the human respiratory syncytial virus, one of the most important respiratory pathogens in young children worldwide. Carvacrol was able to inhibit the virus with a 50% cytotoxic concentration of 250 µg/mL (Pilau et al., 2011). Also, carvacrol was evaluated against *P. aeruginosa*, an opportunistic pathogen implicated in respiratory infections, and showed a 97 and 91% inhibition against this bacteria's adherence and biofilm formation, respectively (Koraichi Saad, Hassan, Ghizlane, Hind, & Adnane, 2011). Ghafari, Sharifi, Ahmadi, and Nayeri Fasaei (2018) reported that oregano EO, rich in carvacrol, showed antibacterial activity (1.25-5.00 µg/mL) against different strains of *S. pneumoniae*, one of the most important causes of respiratory infections, including sinusitis, otitis media, pneumonia, and invasive infections such as septicemia and meningitis.

Today, the pandemic caused by the recently identified respiratory virus SARS-CoV2 has claimed millions of deaths worldwide and is responsible for the recent outbreak of a respiratory infectious disease known as coronavirus disease 19 (COVID19) (Zumla & Niederman, 2020). It is known that the infectious process of SARS-CoV2 begins by the interaction between the spike protein on the surface of the virus and the ACE2 receptors of the human host cells, which allows the virus to enter the cell and the onset of infection (Kulkarni et al., 2020). In this sense, carvacrol received special attention due to recent reports in which their potential to interact with critical molecules of the infectious process was studied by molecular modeling. First, this compound showed the potential to inhibit ACE2 activity, and the authors suggested that it may block the host cell entry of SARS-CoV2 (Abdelli, Hassani, Bekkel Brikci, & Ghalem, 2021). On the other hand, Kulkarni et al. (2020) evidenced by molecular docking the potential of carvacrol to inhibit the binding of viral spike (S) glycoprotein to the host cell. In addition, the modeling reported by Kumar et al. (2020) showed that carvacrol interacts with Mpro, a protease enzyme in the viral genome, which could have a significant effect on the replication and maturation of SARS-CoV2. Therefore, carvacrol can be used as an antimicrobial agent to combat respiratory diseases.

Urinary Tract Microorganisms

Renal infection is a urinary tract infection that usually begins in the urethra or bladder and works up to one or both kidneys. A urinary tract infection is defined as pathogenic microorganisms in any part of the urinary system and represents one of the most common infections worldwide (Bernal-Mercado et al., 2020).

Table 5. Effect of carvacrol against uropathogenic bacteria

Microorganism	Concentration	Results	References
Acinetobacter baumannii	0.16 mg/mL	Inhibited the growth at a subinhibitory concentration	(Montagu et al., 2016)
Uropathogenic *Escherichia coli*	150 and 450 µg/mL	Reduced *E. coli* cell counts in a time-dependent manner caused bacterial membrane disruption, reduced motility, and invasion	(I. Khan, Bahuguna, Kumar, Bajpai, & Kang, 2017)
	150 µg/mL	Reduced cell counts, motility, invasion and induced the release of cellular constituents	(I. Khan et al., 2020)
	<0.01%	Inhibited biofilm formation and decreased the hemagglutinating ability, fimbriae production and the swarming motility	(Lee, Kim, & Lee, 2017)
Enterobacter cloacae	64 and 128 µg/mL	Inhibited biofilm formation and EPS production	(Liu et al., 2021)
Klebsiella pneumoniae	34 µg/mL	Inhibited the growth of all tested *Klebsiella pneumoniae* strains	(Orhan, Ozcelik, Kan, & Kartal, 2011)
Candida albicans	0.1-0.125 mg/mL	Showed fungicidal activity against *Candida* isolates and significant impairment of ergosterol biosynthesis	(Ahmad et al., 2011)
Cryptococcus neoformans	25-102 µg/mL	Inhibited *C. neoformans* growth by increasing fungal membrane permeability	(Nobrega, Teixeira, Oliveira, Lima, & Lima, 2016)

*PVA= polyvinyl alcohol, PVP= polyvinylpyrrolidone, CS= chitosan.

Commonly, uropathogens are fecal contaminants, native microbiota from the skin, or transient microbiota from health staff. The main microorganisms of urinary infections include *Klebsiella*, *Enterobacter*, *Pseudomonas*, *Acinetobacter*, and uropathogenic *E. coli*, the latter being the primary etiological agent. It is well known that the capacity of these microorganisms to adhere to and form biofilms on medical devices are the leading causes of urinary infections associated with catheters (Nicolle, 2014). Commonly, urinary infections are treated with antibiotics; however, bacterial resistance has directed the attention to alternative treatments such as plant extracts, which have been increased in the last years, although its use lies since ancient times. Recent studies evidenced the effect of EOs and their main compounds to threat urinary infections (Ebani, Nardoni, Bertelloni, Pistelli, & Mancianti, 2018; Vamanu, Dinu, Luntraru, & Suciu, 2021). Particularly, carvacrol has been tested against some urinary pathogens, showing a high capacity to inhibit their growth and their colonization in urinary tissues (Table 5).

Antimicrobial Mode of Action of Carvacrol

Carvacrol is a compound recognized for its high antimicrobial activity against different microorganisms. It has been reported that their antimicrobial potential is related to their hydrophobicity.

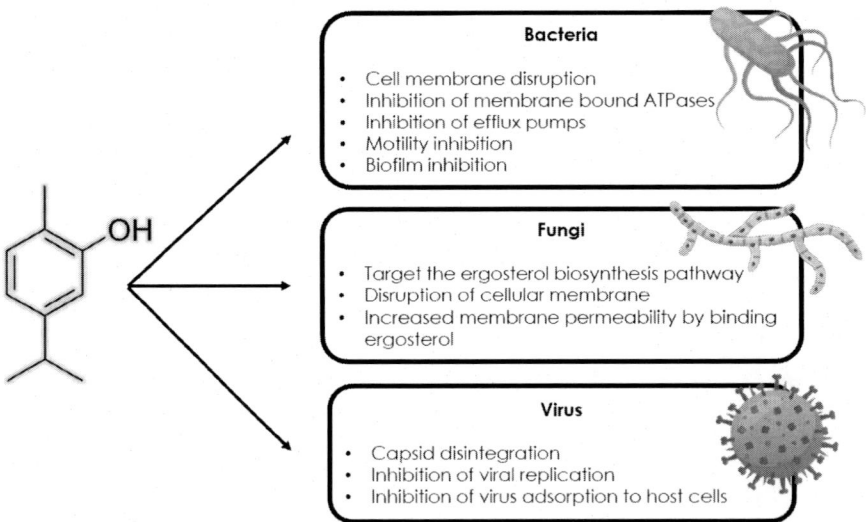

Figure 4. Proposed antimicrobial mechanisms of carvacrol against bacteria, fungi, and viruses.

As explained before, carvacrol has a partition coefficient (LogP) of 3.52, which indicates their affinity to interact with biological membranes (Marchese et al., 2018). Additionally, some authors evidenced that its hydroxyl group (OH) and the presence of a delocalized electron system (double bonds) allows carvacrol to act as a proton exchanger, affecting proton motive force and depletion of the ATP pool, subsequently leading to cell death (Marchese et al., 2018). Different studies have proposed antibacterial (Kachur & Suntres, 2020), antifungal (Ahmad et al., 2011), and antiviral (Gilling, Kitajima, Torrey, & Bright, 2014) action mechanisms such as those mentioned in Figure 4. As evidenced previously, carvacrol possesses high antimicrobial activity against a wide range of microorganisms of medicinal importance.

Antioxidant Properties

Cells naturally produce free radicals during metabolism while producing antioxidants that neutralize these radicals. However, oxidative stress occurs with an imbalance in the cell between free radicals and the decrease in the antioxidant defense (Abdollahi, Ranjbar, Shadnia, Nikfar, & Rezaiee, 2004). Excessive free radicals can damage cell structures such as proteins, lipid membranes, and nucleic acids, leading to cell death. Long-term oxidative stress is associated with the development of numerous conditions such as chronic inflammation, neurodegenerative diseases (Alzheimer's disease and Parkinson's disease), cancer, cardiovascular disorders (high blood pressure, atherosclerosis, and stroke), diabetes, asthma (Maritim, Sanders, & Watkins Iii, 2003; Pohanka, 2014; Reuter, Gupta, Chaturvedi, & Aggarwal, 2010). Free radicals are defined as reactive atoms or molecules with one or more unpaired electrons capable of independent existence (Abdollahi et al., 2004). Free radicals include oxygen/nitrogen species (ROS/RNS) such as superoxide ($O_2^{\cdot-}$), hydroxyl radical ($\cdot OH$), and nitric oxide radical ($\cdot NO$). These molecules are produced as byproducts of mitochondrial metabolism, microbial infections, exercise, and external factors like cigarette smoke, pesticides, ozone, UV radiation, and alcohol (Pisoschi & Pop, 2015).

Antioxidants are stable compounds that can transfer an electron to a free radical and neutralize it, thereby reducing or delaying the free radical's ability to cause damage (Lobo, Patil, Phatak, & Chandra, 2010). Essential oils and their main constituents possess strong antioxidant properties similar to ascorbic acid, vitamin E, and butyl hydroxyl toluene (BHT) (Abou Baker, Al-Moghazy, & ElSayed, 2020; Radünz et al., 2019; Suntres et al., 2015). Remarkably, the antioxidant capacity of carvacrol and its potential to minimize oxidative damage due to its radical scavenging properties are well documented by *in vitro* and *in vivo* studies (M. Khazdair, Alavinezhad, & Boskabady, 2018; Mortazavi et al., 2021). For example, Ibáñez et al. (2020)

reported the antioxidant activity of carvacrol through the DPPH (2,2-diphenyl-1-picrylhydrazyl) method finding that carvacrol has a greater antioxidant capacity than ascorbic acid and the synthetic antioxidant BHT. Also, carvacrol demonstrated a protective effect on chronic stress-induced oxidative stress damage in rat's brain, liver, and kidney by enhancing the activity of antioxidant enzymes (superoxide dismutase, glutathione peroxidase, catalase) and reducing peroxide (H_2O_2), superoxide, and nitric oxide levels (Samarghandian et al., 2016).

Likewise, carvacrol induced a hepatoprotective and antioxidant result by increasing the antioxidant enzymatic activity (superoxide dismutase, catalase, and glutathione peroxidase) and the vitamin C, vitamin E, and reduced glutathione levels in the plasma of rats with D-galactosamine-induced hepatotoxicity (Aristatile, Al-Numair, Veeramani, & Pugalendi, 2009). The study conducted by Mortazavi et al. (2021) reported carvacrol's antioxidant and anti-inflammatory activity on induced liver dysfunction in rats by administering 1 mg/kg of lipopolysaccharides for 2 weeks. The results showed that lipopolysaccharides increased malondialdehyde and nitric oxide levels, decreased superoxide dismutase and catalase activities, causing oxidative stress in treated rats. While the carvacrol administration reversed the effects of exposure to lipopolysaccharides, finding the greater activity of antioxidant enzymes and a decrease in biomarkers of oxidative stress.

Because of its anti-inflammatory and antioxidant properties, Ghorani et al. (2021) proposed using carvacrol as an asthma treatment. In asthmatic patients, treatment with carvacrol (1.2 mg/kg/day for 2 months) reduced respiratory symptoms and improved oxidative stress indicators and anti-inflammatory cytokine levels in serum. A clinical trial randomized with 20 humans exposed to sulfur mustard to induce lung injury demonstrated that carvacrol treatment (1.2 mg/kg/day) for 2 months reduced inflammatory cells and oxidant biomarkers (malondialdehyde) while increasing antioxidant biomarkers (superoxide dismutase and catalase) and improving pulmonary function tests (M. Khazdair et al., 2018). Similarly, Önal et al. (2011) demonstrated the protective effect of carvacrol (73 mg/kg) on lung tissue of Wistar rats exposed to oleic acid, which generates acute lung damage with symptoms very similar to those presented in human patients. The results showed that carvacrol significantly decreased the concentration of malonaldehyde, a product of lipid peroxidation, in lung tissue. The studies above suggest that the antioxidant effect of carvacrol could prevent pulmonary tissue damage.

The oral administration of carvacrol at 25, 50, and 100 mg/kg for 40 days ameliorated the learning deterioration, memory capacities, neurodegeneration, antioxidant capacity, and lipid peroxidation in the hippocampus of Wistar rats exposed to lead acetate (Mehrjerdi et al., 2020). Similarly, Deng, Lu, and Teng (2013) demonstrated that carvacrol (25, 50, and 100 mg/kg for 7 weeks) returned the contents of malonaldehyde, superoxide dismutase, and glutathione, related to oxidative stress, towards their control values suggesting the reduction of diabetes-associated cognitive deficit in diabetic rats. On the other hand, it has been suggested that carvacrol could be a potent candidate for treating acute pancreatitis via its antioxidative mechanism of action. For example, carvacrol (50, 100, and 200 mg/kg) provided hepatic protection against cerulein-induced acute pancreatitis in rats (Bakır et al., 2016). Carvacrol at higher doses reduced malondialdehyde and 8-hydroxydeoxyguanosine, a marker of DNA oxidative damage. Furthermore, carvacrol increased the liver superoxide dismutase, catalase, and glutathione peroxidase activities and alleviated necrosis, coagulation, and inflammation. Similar results were reported by Kılıç et al. (2016), who showed that carvacrol reduced cell injury and modulated oxidative stress in the pancreas of rats with acute pancreatitis.

In diabetic Wistar male rats, treatment with carvacrol at a dose of 75 mg/kg for 8 weeks improved the histological morphology of the testis, lowered tissue activity of superoxide dismutase and glutathione peroxidase enzymes, and decreased malondialdehyde levels (Shoorei et al., 2019). This therapy also had anti-proliferative effects, as it reduced Bax and increased *bcl-2* gene and protein expression and mitigates testicular tissue damage reducing germ cell apoptosis. Similarly, the monotherapy of carvacrol and its isomer thymol (10 and 20 mg/kg) reduced the oxidative damage, elevated the antioxidant levels, and enhanced the sperm quality parameters in rats' models (Güvenç, Cellat, Gökçek, Yavaş, & Yurdagül Özsoy, 2019). Both terpenoids decreased malondialdehyde levels in testicles, liver, and kidney tissues and increased spermatozoa concentration and motility, but only thymol increased glutathione levels, and carvacrol increased glutathione peroxidase and catalase activity.

Carvacrol antioxidant mode of action is based on its structure and is related to the donation of hydrogen atoms to unpaired electrons, becoming a radical that is stabilized by its resonance structure (Figure 5). The benzene ring has a hydroxyl substituent at C5 (meta-position), and since the bond between oxygen and benzene is very stable, the hydroxyl group can be easily deprotonated. In monophenols, the ability to give up H^+ is associated with reducing free radicals, stabilizing oxygen singles or triplets, and chelating action on transition metals (Peralta-Pérez & Volke-Sepúlveda, 2012; Rodriguez-Garcia et al., 2016). All the reviewed studies reaffirm the potential of carvacrol as an antioxidant agent, whose administration could help reverse or prevent various pathologies.

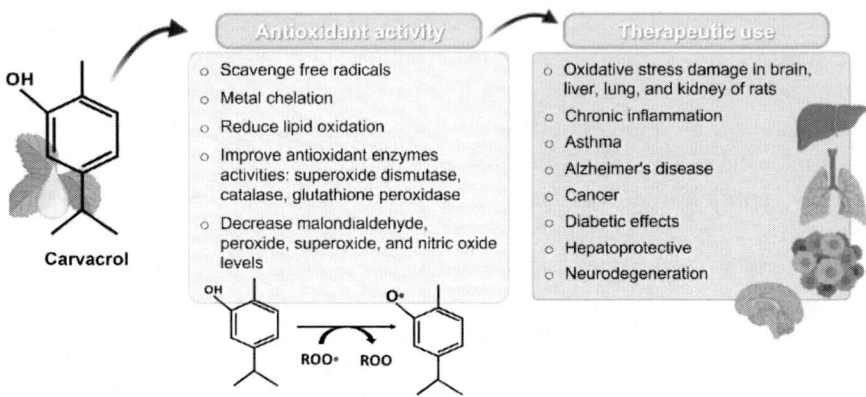

Figure 5. Carvacrol antioxidant activity and its promising therapeutic use.

Anti-Inflammatory Activity

The inflammatory process forms an essential part of the immune response as a defense mechanism for pathogenic microorganisms' infection or exposure to chemical agents harmful to the body (Rock & Kono, 2008). However, when this response is exaggerated and lacking in regulation, it could cause tissue injury, physiological decompensation, organ dysfunction, and death (Sherwood & Toliver-Kinsky, 2004). The inflammatory response involves different inflammatory leukocytes cells like macrophages, neutrophils, and lymphocytes, which release specialized substances called inflammatory mediators (Abdulkhaleq et al., 2018). Also, various chemical mediators are released from the circulation system and injured tissue (Halliwell & Gutteridge, 2015). Some molecule mediators include histamine, serotonin, prostaglandins, leukotrienes, oxygen- and nitrogen-derived free radicals, bradykinin, proinflammatory cytokines (interleukin (IL)-1β, IL-8, tumor necrosis factor-alpha (TNF-α), IL-6, and IL-12), and others which have a significant role in regulating the immune system (Abdulkhaleq et al., 2018). These substances cause the dilation of small blood vessels, increasing blood flow, permeability, the migration of fluid, proteins, and more immune system cells from circulation to the injured tissue (Encyclopedia-Britannica, 2020). All these mechanisms cause harmed areas to swell, turn red, feel hot and painful. The inflammation process can be classified into acute and chronic responses. Acute inflammation is characterized by a relatively short duration (hours to days) and involves the previous mechanism described. Chronic inflammation is referred to a more prolonged response (months to years) and includes some human diseases such as rheumatoid arthritis, Crohn's disease, tuberculosis, and chronic lung disease (Sherwood & Toliver-Kinsky, 2004). Severe neurodegenerative diseases such as Alzheimer's or Parkinson's can be worsened by the deregulated secretion of inflammatory mediators (Gupta, Kunnumakkara, Aggarwal, & Aggarwal, 2018; Rea et al., 2018).

The use of alternative therapies based on natural compounds is a promising treatment to alleviate the complications caused by the inflammatory process of many diseases (Jalalvand, Shahsavari, Sheikhian, Ganji, & Mosayebi, 2020). Compounds of natural origin such as carvacrol have demonstrated an anti-inflammatory activity due to their ability to suppress the expression of cyclooxygenase (COX)-2, inhibit the production and actions of nitric oxide (NO), and inhibit the signaling cascade that causes the synthesis and release of proinflammatory cytokines in affected tissues (Figure 6) (Guimarães et al., 2012; Hotta et al., 2010). For example, de Carvalho et al. (2020) reported the anti-inflammatory effect of carvacrol by inhibiting COX-2, an essential enzyme responsible for transforming arachidonic acid into prostaglandins and leukotrienes involved in the inflammatory process. In addition, this terpenoid decreased the prostaglandin E2 (PGE2) concentration and the proinflammatory interleukin IL-8 levels, which participate in the immune system by the activation of neutrophils, increment of adhesion proteins in endothelial cells, and oxidative metabolism by polymorphonuclear neutrophils (de Carvalho et al., 2020).

In the same approach, in a study conducted by Landa, Kokoska, Pribylova, Vanek, and Marsik (2009), carvacrol in a range of 0.1 – 100 μM inhibited the COX-2 and COX-2 isoform activity *in vitro* and reduced the PGE2 levels. Jalalvand et al. (2020) carried out a comparative study of the anti-inflammatory effect of *Satureja khuzestanica* EO and carvacrol on a macrophage cell line. The results showed that although the EO has a more significant anti-inflammatory effect, carvacrol by itself inhibits the expression of the *cox-2* gene that codes for the synthesis of COX-2. This enzyme has an essential role in Parkinson's development; thus, the results of this study suggest that the reduction in this enzyme activity can generate an improvement in the disease. Likewise, da Silva Lima et al. (2013) found that carvacrol (50 and 100 mg/kg) attenuated mice's paw edema induced by reducing IL-1β and PGE2 levels and COX-2 and IL-1β mRNA expression. Also, this treatment increased IL-10 (anti-inflammatory cytokine) levels and mRNA expression in the inflamed paw. Moreover, carvacrol supplementation (1.2 mg/kg/day for two months) in patients with lung lesions caused a decrease in proinflammatory cytokines IL-2, IL-4, IL-10, IL-8, IL-6, and TNF-α and an increase in anti-inflammatory cytokines IL-10 and interferon-gamma (IFNγ) levels (M. R. Khazdair & Boskabady, 2019).

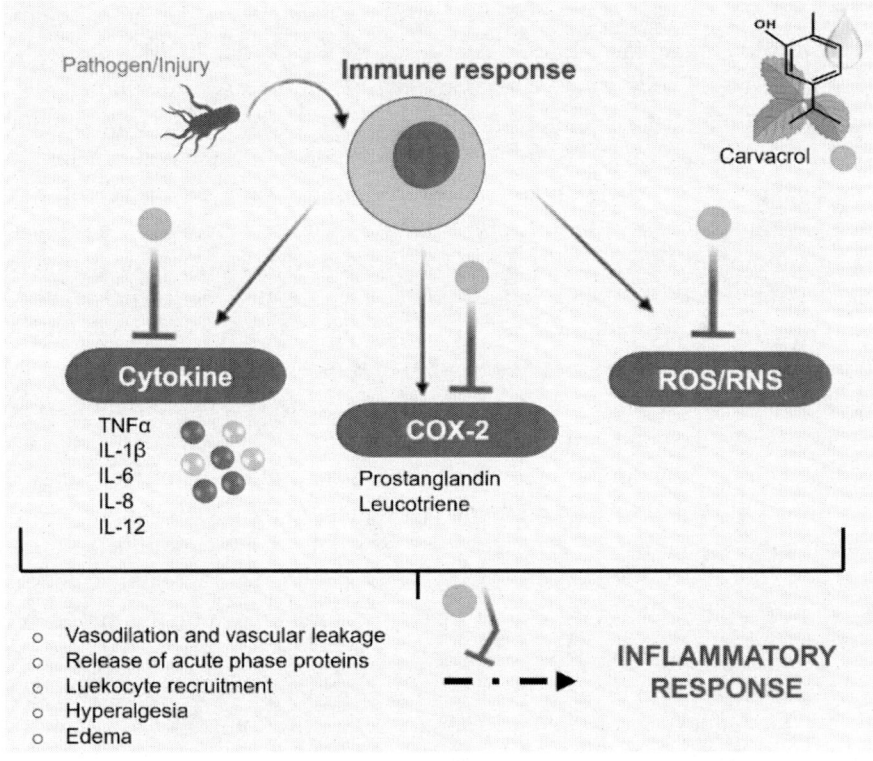

Figure 6. General anti-inflammatory activity of carvacrol.

Carvacrol (20, 40 or 80 mg/kg) attenuated lipopolysaccharide-induced endotoxemia and acute lung injury in mice by the inhibition of proinflammatory cytokines (TNF-α, IL-6, and IL-1β) production and nuclear factor-κB (NF-κB) and mitogen-activated protein kinase (MAPKs) cascade activation (Feng & Jia, 2014). These results suggested the potential use of carvacrol to reduce severe cases of acute respiratory distress syndrome and acute lung lesions development. Also, asthma is a chronic disease that affects many people worldwide. It is associated with inflammation of the bronchi and whose progression is mediated by an inflammatory response initiated by increased cytokines and interleukins (Ezz-Eldin et al., 2020). In this context, Ezz-Eldin et al., (2020) reported that carvacrol administration in bronchial asthma-induced rats reduced interleukins levels such as IL-13, IL-5, IL-4, IgE, TNF-α, and among other proinflammatory mediators. The suggested mechanism of action shows that carvacrol is a potent suppressor of eosinophils, which are the primary source of IL-5. This, together with a regulation in the balance of Th1 and Th2 cytokines, demonstrates carvacrol's potential as an antiasthmatic treatment by limiting the cascade of inflammatory cytokines and cell adhesion molecules.

On the other hand, multiple sclerosis development is associated with the inflammatory process (Mahmoodi et al., 2019). Therefore, the anti-inflammatory effect of carvacrol was studied in experimental autoimmune encephalomyelitis in mice models. Carvacrol administration (5 and 10 mg/kg) ameliorated clinical signs, modulated pro- and anti-inflammatory cytokines, and reduced leukocytes infiltration into the central nervous system (Mahmoodi et al., 2019). Cancer is another disease in which inflammatory processes play a crucial role in the symptoms and development of severe cases (Aravindaram & Yang, 2010). The increase in reactive oxygen species due to the anaerobic conditions created inside solid tumors such as colon cancer causes symptoms that deteriorate patients' quality of life and nutritional status. Arigesavan and Sudhandiran (2015) conducted a study in rats with induced colon cancer by applying 1,2 dimethylhydrazine (DMH) and sodium dextran sulfate (DSS) together with the administration of carvacrol before and after disease induction. An increase in the activity of antioxidant enzymes such as catalase and superoxide dismutase and a decrease in the concentration of peroxidized lipids and reduced glutathione were found. In rats treated with carvacrol, they also reported suppressing inflammation mediators such as nitric oxide synthase and interleukin IL-1β. Thus, the authors suggest that the administration of carvacrol as a dietary antioxidant and an anti-inflammatory agent can be used as a novel treatment for colon cancer developed by chronic inflammation. The above studies suggest the promising use of carvacrol in anti-inflammatory therapy, however more *in vivo* studies are required.

Anti-Cancer Activity

Cancer is one of the leading causes of mortality worldwide, reaching about 19.3 million new cases and 9.9 million deaths in 2020 (GLOBOCAN, 2020). It is estimated that in 2040, cancer cases will achieve 29.5 million diagnostics and 16.4 million deaths (IARC, 2018). Cancer is a multifactorial pathology that involves mainly acquired genetic abnormalities by the exposition to exogenous and endogenous agents (90 – 95%) as well as with inherited genetic aberrations (5-10%) (Maru, Hudlikar, Kumar, Gandhi, & Mahimkar, 2016). Cancer develops through three stages: initiation, promotion, and progression (Maru et al., 2016). Among these, the two early stages are critical to be prevented or suppressed through anticarcinogens (blocking agents) (Wu, Patterson, & Hawk, 2011). Each phase involves many cell alterations; thus, different chemopreventive and chemotherapeutic strategies are necessary (Blowman, Magalhães, Lemos, Cabral, & Pires, 2018). Initially, cancer can be prevented with the use of antimutagenic compounds to block the initiation process that is characterized by DNA damage. On the other hand, when the mutation cannot be repaired, it is necessary the use of antiproliferative compounds that can inhibit the promotion stage of carcinogenesis. This stage is reversible and its related to an out of control' proliferation attributed to mutations in proto-oncogenes and/or inactivation of tumor suppressor genes responsible to the normal cell growth (ACS, 2014).

Traditional therapy to treat cancer is not entirely effective, exerts strong side effects attributed to the non-selectivity action on cells (cancerous and non-cancerous), shows a higher rate of multidrug resistance, and is highly expensive (Fekrazad et al., 2017). Therefore, much research is currently conducted to search for new anti-cancer compounds that can prevent mutation in cells, inhibit cancer cell proliferation, and induce apoptosis. In particular, oregano EO has exerted anti-cancer activity by mechanisms such as antioxidant, antimutagenic and antiproliferative activities (Dutra et al., 2019) mainly attributed to carvacrol, its principal constituent (Karadayi, Yildirim, & Güllüce). Although carvacrol has proved a wide range of bioactivities, scarce studies are reported regarding the antimutagenic and antiproliferative activity.

Mezzoug et al. (2007) evaluated the capacity of carvacrol (0.2%) to inhibit the mutagenicity of the indirect mutagen urethane (URE) at 10 mM through the SMART assay (Somatic Mutation and Recombination Test) in *Drosophila melanogaster*. Carvacrol suppressed the URE-induced spots mutations by 63% associated with the antioxidant activity that could scavenge the electrophile metabolites of urethane after bioactivation. Similarly, Ipek et al.

(2005) demonstrated that carvacrol could act as a strong direct mutagen (0.01-1 μg/plate) against nitro-*O*-phenylenediamine (200 μg/plate) and 2-aminofluorene (1 μg/plate), suggesting the importance of carvacrol for the prevention of cancer. The assay was conducted with *Salmonella* Typhimurium TA98 and TA100 strains in the Ames *Salmonella*/microsomal test with and without S9 metabolic activation.

In the same field, Bound, Murthy, Negi, and Srinivas (2020) reported that carvacrol (0.5 and 1 μmol/plate) exerted a strong antimutagenicity showing a 61 and 82.3% inhibition, respectively, of the mutagen methyl methanesulfonate (1.5 μg/plate) in *S*. Typhimurium TA98 and TA1538 strains. The effects observed were related to a possible interaction with the mutagen, preventing their entrance into the cell or induction of detoxifying enzymes. Also, carvacrol (0.05 mM) significantly reduced the oxidative DNA damage induced by the heterocyclic aromatic amine 2-methimidazo[4,5-f]-quinoline (IQ), hydrogen peroxide, and mitomycin C in human lymphocytes. However, the results showed that carvacrol at higher concentrations than 0.1 mM might cause DNA damage (Aydın, Başaran, & Başaran, 2005a, 2005b). Oezbek et al. (2008) evaluated the antimutagenic effect of oregano EO (0.05, 0.5, 5 μg/plate) on the direct-acting mutagens 4-nitro-1-quinoline oxide (4NQO) and sodium azide (NaN3) and the indirect mutagen 2-aminofluorene (2-AF) on *Salmonella typhimurium* TA1535 and TA1538 strains. The carvacrol antimutagenic effect on non- and dependent bioactivation mutagens was related to the reactive oxygen species (ROS) scavenging potential and the inhibition of CYP450 enzymes (mutagen activation), respectively.

The mechanisms involved to protect against cell mutation are the interaction with the mutagen (direct or indirect mutagen), inhibition of the catalytic activity of the xenobiotic-metabolizing enzymes, scavenging of metabolites formed after bioactivation, and induction of DNA repairing mechanisms (Słoczyńska, Powroźnik, Pękala, & Waszkielewicz, 2014) (Figure 7). However, the exact mechanisms for natural compounds are not fully understood.

Additional to the antimutagenic effect, several studies have demonstrated the promising effect of carvacrol in the second stage of carcinogenesis, inducing antiproliferative and apoptotic activities against cancer cells (Fan et al., 2015; Jayakumar et al., 2012; Jung, Kim, & Lee, 2018). The exact molecular mechanisms implicated in the anti-cancer activity of carvacrol are not completely defined. The anti-cancer mechanisms reported now involve disruptions of mitochondrial membrane potential, formation of ROS, expression, and repression of specific proteins of the apoptotic cascade observed in different organ-cancer cell lines (Figure 8) (Arunasree, 2010; Günes-Bayir et al., 2017). For example, carvacrol exerted an antiproliferative activity on lung human cancer cells A549 and H460 by reducing AXL (receptor tyrosine kinase) protein level. AXL is responsible for transducing extracellular signals to promote cell survival, proliferation, and inhibition of apoptosis in cancer cells (Jung et al., 2018).

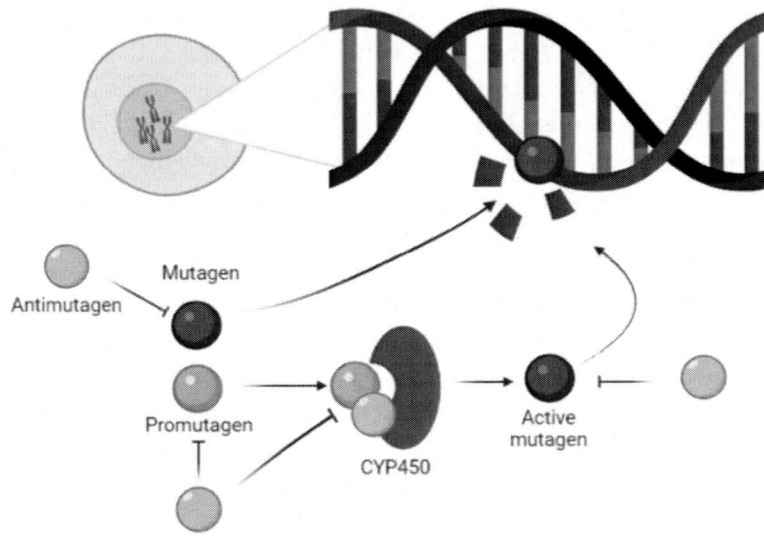

Figure 7. Possible carvacrol antimutagenic mechanisms of action.

Similarly, Fan et al. (2015) demonstrated that carvacrol (200-900 μmol/L) inhibited the proliferation of colon cancer cells HCT116 and LoVo by decreasing the expression of two matrices metalloproteinases (MMP-2 and MMP-9). Also, it induced the cell cycle arrest by blocking the G2/M phase and reducing the cyclin B expression. Mastelic et al. (2008) reported an antiproliferative activity of carvacrol against human cervical carcinoma cells (HeLa) in a dose-dependent manner (0.1-10 mM), decreasing mitochondrial activity. Carvacrol exerted an antiproliferative activity on human gastric adenocarcinoma cells (IC_{50}= 82.57 ± 5.58 μmol/L) attributed to an antioxidant and apoptotic effect associated with the increase of initiator (*caspase-9*) and effector (*caspase-3*) proteins of the apoptosis cascade expression; as well as the increase of pro-apoptotic protein Bax (Günes-Bayir et al., 2017). On the contrary, Bcl-2 protein density decreased dose-dependent, suggesting that the apoptotic effect was also related to reducing this inhibitory protein of apoptosis (Günes-Bayir et al., 2017).

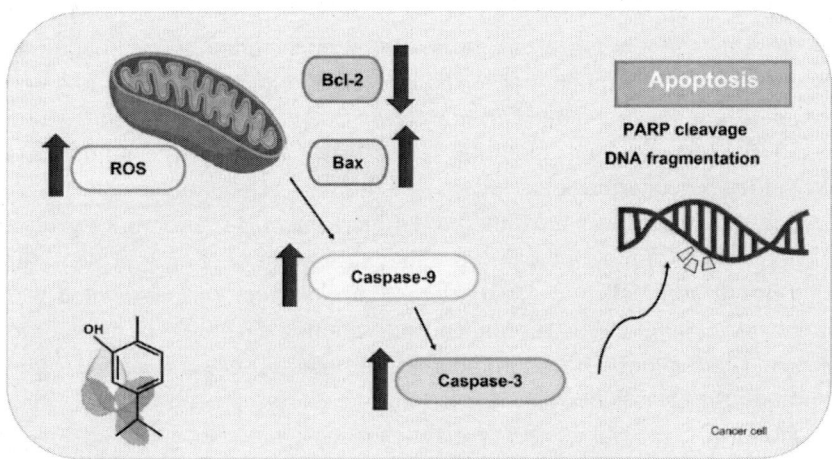

Figure 8. General antiproliferative mechanisms of carvacrol.

In the same approach, Heidarian and Keloushadi (2019) reported that carvacrol (360 µM) reduced the proliferation of PC3 prostate cancer cells via reduction of *il-6* gene expression and diminished the signaling proteins pSTAT3, pAKT3, and pERK1/2 associated with cell proliferation. Similarly, carvacrol induced apoptosis in human prostate cancer cells via ROS and cell cycle arrest at G0/G1 (F. Khan et al., 2017). Carvacrol also showed an antiproliferative effect in metastatic breast cancer cells (MDA-MB231) associated with activation of apoptosis response and a decrease in mitochondrial membrane potential resulting in cytochrome C release from mitochondria, caspase activation, cleavage of PARP, and fragmentation of DNA (Arunasree, 2010; Yin et al., 2012). Al-Fatlawi (2018) described the anti-proliferative and anti-apoptotic effect of carvacrol (40-450 µM) on human cervical cancer cells (SiHa) and human liver cancer cells (HepG2) in a dose-dependent manner in the MTT (3-(4, 5-dimethyl-2-thiazolyl)-2, 5-diphenyl-2H tetrazolium bromide) and LDH (lactate dehydrogenase enzyme) assays. These results showed that carvacrol induces apoptosis regulatory genes in these cancer cell lines, thus reducing the proliferation of cells.

Carvacrol exhibited antiproliferative and apoptotic activity against choriocarcinoma cell lines (JAR, JEG3) through multiple mechanisms of action (Lim, Ham, Bazer, & Song, 2019). These mechanisms involved an increase in the expression of the pro-apoptotic proteins Bax, Bak, and cytochrome c in a dose-dependent manner (0, 100, 200, and 300 µM) and no changes in the expression of Bcl-xL antiapoptotic protein. In this study, carvacrol also suppressed the signaling proteins PI3K–AKT and MAPK. It affected the mitochondrial membrane potential inducing a liberation of calcium ions from mitochondria and generated oxidative stress and lipidic peroxidation in the evaluated cell lines (Lim et al., 2019). Bhakkiyalakshmi et al. (2016) reported an apoptotic effect of carvacrol on human acute promyelocytic leukemia cells (HL-60) and human T lymphocyte cells (Jurkat). The cytotoxic effect was attributed to an increase in Bax protein's expression (pro-apoptotic) and a decrease in the expression of Bcl-2 protein (anti-apoptotic) in both cell lines. Also, the apoptotic effect was associated with a mitochondrial potential disruption and activation of caspase-3 of the apoptosis cascade. The depletion of the mitochondrial potential was associated with the suppression of Bcl-2 because these family proteins are responsible for the mitochondrial potential's stability maintenance.

Limited in vivo studies have reported the antiproliferative effect of carvacrol. For example, Jayakumar et al. (2012) investigated the carvacrol effect on diethylnitrosamine-induced liver cancer in male Wistar albino rats. This study confirms the carvacrol's potential use as a chemopreventive during liver cancer progression. Furthermore, the results revealed that carvacrol supplementation (15 mg/kg body weight) attenuated liver foci and nodules' appearance and showed free radical scavenging and antioxidant activities to modulate lipid peroxidation levels and increased the endogenous antioxidant mechanisms in induced hepatocellular carcinogenesis. All the studies above demonstrated the potential of carvacrol as an anti-cancer agent to prevent the initiation stage of carcinogenesis and suppress the proliferation of cancerous cells. Thus, carvacrol may be a helpful alternative to chemoprevention against several human cancers and as a therapeutic agent.

Strategies to Improve the Biological Effects of Carvacrol

As stated before, carvacrol presents a wide range of biological properties (Suntres et al., 2015). However, carvacrol has limited application in the pharmaceutical area due to its low water solubility, volatile nature, poor bioavailability, hydrophobic and irritant characteristics (Shinde et al., 2020). For this reason, new strategies have been sought to improve the effectiveness of carvacrol, either using the combination with other compounds, finding transporters to be taken to specific sites and drug-controlled release systems (Figure 9) (Niaz et al., 2021; Shinde et al., 2020; Trindade et al., 2019).

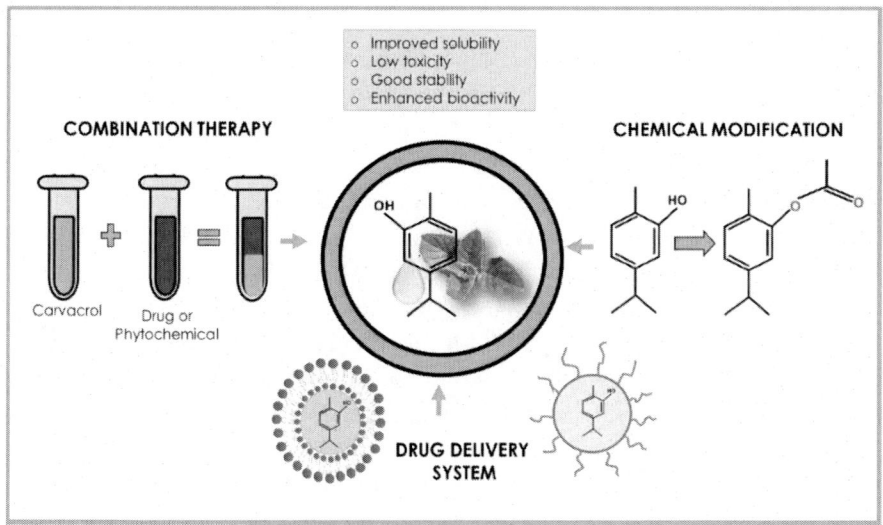

Figure 9. Strategies to improve the biological effects of carvacrol.

Carvacrol Activity in Combined Therapy

Several studies have demonstrated beneficial biological effects when combining carvacrol with other phytochemicals or with drugs. For example, carvacrol, cinnamaldehyde, and honey combined showed a synergistic effect against 77.3% of *P. aeruginosa* isolated from burn wound infections (Mohammadzamani et al., 2020). The results revealed that the MIC of the combination was 100 times lower than imipenem, a current antibiotic used to treat *P. aeruginosa*. Moreover, this combination decreased the expression of some virulence genes (*exoS* and *ampC*). Therefore, the combination of carvacrol with other compounds is a promising alternative to treat bacterial infections. Similarly, in search of new therapeutic strategies to treat acute myeloid leukemia, the combination of carvacrol and thymol from *Ptychotis verticallata* EO has been suggested as an effective therapy against cancer cell lines (Bouhtit et al., 2021). This study demonstrated that the synergistic combination could protect normal cells from drug toxicity and eradicate leukemic cells (KG1 and HL60) by different molecular pathways such as apoptosis, oxidative stress, reticular, autophagy, and necrosis.

The combined therapy of thymol (20 mg/kg p.o.) and carvacrol (25 mg/kg p.o.) for 14 days before and 2 days after doxorubicin administration (to induce cardiotoxicity in male rats) showed a cardioprotective synergistic effect ameliorating heart function and oxidative stress parameters attributed to antioxidant, anti-inflammatory, and anti-apoptotic activities (El-Sayed, Mansour, & Abdul-Hameed, 2016). Likewise, the combination of carvacrol and thymol protected against $A\beta_{25-35}$-induced cytotoxicity in PC12 cells related to the brain senile plaque in Alzheimer´s disease. Both compounds reversed the produced intracellular ROS and increased the protein kinase C activity as memory-related protein, suggesting potential in modulating Alzheimer´s disease (Azizi, Salimi, Amanzadeh, Majelssi, & Naghdi, 2020).

In a similar approach, carvacrol combined with erythromycin showed a synergistic effect against erythromycin-resistant *S. pyogenes* Group A, especially in those strains where erythromycin resistance is only expressed in the presence of antibiotic sub-inhibitory concentration (Magi, Marini, & Facinelli, 2015). The combination of carvacrol and citral enhance the antibiotic activity of erythromycin, bacitracin, and colistin by reducing MIC values against *L. monocytogenes* and *L. innocua* (Zanini, Silva-Angulo, Rosenthal, Rodrigo, & Martínez, 2014). The authors suggested that carvacrol could disrupt bacterial membrane and allow the nonspecifically antibiotic entrance to bacterial cells. The combination of carvacrol with other bioactive compounds or pharmaceutical drugs can potentiate the activity by adding different modes of action. This approach has many benefits: reduction of dose-related toxicity, a greater spectrum of activity, and prevention of drug resistance (Zanini et al., 2014).

Carvacrol Chemical Modification and Prodrugs Design

The chemical modification or synthesis of compound derivatives are promising strategies to overcome physiological barriers and improve the physicochemical properties of carvacrol (Marinelli et al., 2018). In this sense, Bassanetti et al. (2017) synthesized sulfonic, sulfonic potassium salt, and laurate ester derivatives of carvacrol with reduced volatility, higher water solubility, and non-cytotoxicity towards HT-29 human cells. Furthermore, these compounds showed antimicrobial activity against *E. coli, S.* Typhimurium, *S. enteritidis,* and *C. perfringens* highlighting the correlation between the free-hydroxyl group in the phenolic ring and the antimicrobial potency. Similarly, carvacryl acetate, a semisynthetic carvacrol derivative, was explored as an anti-inflammatory and anti-nociceptive agent in mice (Damasceno et al., 2014). The derivative reduced carrageenan-induced paw edema and edema induced by inflammatory mediators: histamine, serotonin, prostaglandin E2. The results showed that carvacryl acetate increased anti-inflammatory IL-10 levels and reduced mas cell degranulation, neutrophil migration, myeloperoxidase, and proinflammatory cytokine IL-1β levels. Also, it presented an anti-nociceptive effect via capsaicin and glutamate pathways.

The design of codrugs or prodrugs is an effective strategy to ameliorate the carvacrol physicochemical properties with a potentiated efficacy due to the synergistic mode of action of the molecules involved (Cacciatore et al., 2015). Prodrugs are molecules that, after administration, are biotransformed by chemical or enzymatic hydrolysis, releasing the parent molecule (Jornada et al., 2016). Marinelli et al. (2019) designed 23 different carvacrol prodrugs: amino-acid ester prodrugs to enhance carvacrol solubility and prenylated prodrugs to improve its lipophilicity, affinity, and interaction with bacterial membrane. The most effective carvacrol prodrugs used prenylated chains (isopentenyl, geranyl, and farnesyl) directly conjugated to the carvacrol hydroxyl group through ether junction. These prodrugs showed good antibacterial and antibiofilm effects against *S. aureus* and *S. epidermis*. Also, they were more stable in simulated fluids and human plasma and showed no toxicity in human erythrocytes and HaCaT cells.

In the same approach, de Oliveira Pedrosa Rolim et al. (2019) synthesized a codrug developed by carvacrol with ibuprofen, generating a new hybrid chemical entity to search for new drugs with anti-inflammatory properties. This codrug showed good oral bioavailability, gastrointestinal tract absorption, and low toxicity. In addition, *in vitro* experiments showed that the hybrid reduced the levels of IL-2, IL-10, IL-17, cytokines, and INF-Y, while *in vivo* trials showed a decrease in total white blood cell count and IL-1β levels. These results indicated that the codrug is an efficient anti-inflammatory agent. On the other hand, Cacciatore et al. (2015) developed a carvacrol derivative by conjugating the carvacrol hydroxyl group to the carboxyl portion of sulfur-containing amino acids through an ester bond. The derivative was no toxic and exhibited antibacterial and antibiofilm activities against *S. aureus, E. coli,* and *C. albicans*. Also, it presented low water solubility, good stability under acidic conditions, in the presence of pepsin and pancreatin, in plasma of human and rat, and it showed good permeability measured by parallel artificial gastrointestinal membrane permeability assay. All these results suggest the possible cross of stomach environment without modifications, the absorption in the gastrointestinal tract after oral administration and the release of carvacrol after enzymatic hydrolysis.

Carvacrol Delivery Systems

Drug delivery systems using different nanomaterials can enhance the solubility, controlled release, and potential activities of carvacrol (Marchese et al., 2018). In this context, Eusepi et al. (2020) proposed a release system for carvacrol prodrugs using Palygorskite and Sepiolite nanocarriers to improve their pharmacokinetic and biopharmaceutic profiles. The results showed that the mineral clay nanocarriers could stabilize prodrugs and ensure a sustained release of carvacrol prodrugs that underwent a conversion into active carvacrol with antimicrobial properties. This system of releasing carvacrol prodrugs using nanocarriers should improve bioavailability in the small intestine. Similarly, Mir et al. (2020) produced a carvacrol-poly(caprolactone) nanoparticle delivery system to achieve an antimicrobial effect at specific chronic wound infections sites. The nanoparticles systems were incorporated into solvent microneedles to facilitate their application at the specific site of necrotized tissue. This study demonstrated that encapsulation of carvacrol in caprolactone increased its antimicrobial activity by 2 to 4 times against *S. aureus* and *P. aeruginosa* and improved the skin retention of carvacrol after 24 h (Mir et al., 2020). In the same approach, the study conducted by Shakeri et al. (2019) showed the preparation and optimization of solid beeswax lipid nanoparticles loaded with carvacrol and astaxanthin. This nano-system presented a high entrapment efficacy, good compound release, great stability under acidic and alkaline conditions, significant radical scavenging activity, and efficient removal and killer agent against *P. aeruginosa* and *S. aureus* biofilms cells.

To improve the oral delivery, solubility, bioavailability, and efficacy of carvacrol, Shinde et al. (2020) incorporated carvacrol into lecithin-stabilized zein nanoparticles as a promising strategy for colon cancer treatment. This system showed good encapsulation efficiency and a controlled release in 24 h. Also, the nanoparticles increased antioxidant activity and induced cytotoxicity against colon cancer (SW480) cells. Moreover, in need of new alternatives to treat inflammatory muscle injury, Souza et al. (2018) developed an inclusion complex of β-cyclodextrin (βCD) and carvacrol to reduce acute skeletal muscle inflammation in rats models. The βCD-carvacrol (80 or 160 mg/kg) orally administered in rats reduced the myeloperoxidase activity, edema, and mechanical hyperalgesia 24 h after injury. Also, the complex showed anti-inflammatory effects, reducing IL-1β, IL-6, and MIP-2 levels and increasing IL-10 six hours after induction compared to the control group. With all these strategies to improve the biological effects of carvacrol, there is still research aimed at improving these effects to obtain a compound as applicable as possible in the pharmaceutical industry.

Conclusion

This review presents the potential use of carvacrol in clinical treatments. Carvacrol possesses various biological and therapeutic properties such as antioxidant, antimicrobial, anti-inflammatory, and anti-cancer activities. Carvacrol can inhibit viruses, bacteria, and fungi growth, including *E. coli, Salmonella, L. monocytogenes, S. aureus, P. aeruginosa, Saccharomyces,* and *Candida,* and it can affect some virulence factors like motility, biofilm formation, and *quorum sensing*. Moreover, it is a potent antioxidant and anti-inflammatory agent to help prevent certain diseases. In addition, much research has demonstrated that carvacrol exhibits anti-cancer activity through apoptosis and inhibition of proliferation mechanisms. However, the therapeutic application of carvacrol may be comprised, and some strategies such as therapy combination, chemical modification and controlled drug release systems have been developed to improve carvacrol solubility and bioavailability in biological tissues. These studies seem to be promising to produce alternative medicines using carvacrol that can be helpful in therapeutic applications. Although there is enough literature on the carvacrol therapeutic potential, there is still much more to investigate before the molecule can be used as a clinical drug. Most of these studies have been carried out *in vitro,* and few use animal models; hence, further research must be addressed on the carvacrol clinical use in humans to validate the *in vitro* results and evaluate its bioavailability effects and mechanisms.

References

Abdelli, I., Hassani, F., Bekkel Brikci, S., and Ghalem, S. (2021). In silico study the inhibition of angiotensin converting enzyme 2 receptor of COVID-19 by *Ammoides verticillata* components harvested from Western Algeria. *Journal of Biomolecular Structure and Dynamics, 39*(9), 3263-3276.

Abdollahi, M., Ranjbar, A., Shadnia, S., Nikfar, S., and Rezaiee, A. (2004). Pesticides and oxidative stress: a review. *Medical Science Monitor, 10*(6), RA141-RA147.

Abdulkhaleq, L., Assi, M., Abdullah, R., Zamri-Saad, M., Taufiq-Yap, Y., and Hezmee, M. (2018). The crucial roles of inflammatory mediators in inflammation: A review. *Veterinary World, 11*(5), 627.

Abou Baker, D., Al-Moghazy, M., and ElSayed, A. (2020). The in vitro cytotoxicity, antioxidant and antibacterial potential of *Satureja hortensis* L. essential oil cultivated in Egypt. *Bioorganic Chemistry, 95*, 103559.

ACS. (2014). *Oncogenes and Tumor Suppresor Genes*. Retrieved from https://www.cancer.org/cancer/cancer-causes/genetics/genes-and-cancer/oncogenes-tumor-suppressor-genes.html.

Ahmad, A., Khan, A., Akhtar, F., Yousuf, S., Xess, I., Khan, L., and Manzoor, N. (2011). Fungicidal activity of thymol and carvacrol by disrupting ergosterol biosynthesis and membrane integrity against *Candida*. *European Journal of Clinical Microbiology and Infectious Diseases, 30*(1), 41-50.

Ahmadi, S., Fazilati, M., Nazem, H., and Mousavi, S. M. (2021). Green synthesis of magnetic nanoparticles using *Satureja hortensis* essential oil toward superior antibacterial/fungal and anticancer performance. *BioMed Research International, 2021*, 1-14.

Ait-Ouazzou, A., Espina, L., García-Gonzalo, D., and Pagán, R. (2013). Synergistic combination of physical treatments and carvacrol for *Escherichia coli* O157:H7 inactivation in apple, mango, orange, and tomato juices. *Food Control, 32*(1), 159-167. doi:https://doi.org/10.1016/j.foodcont.2012.11.036.

Al-Fatlawi, A. (2018). Anti-proliferative and pro-apoptotic activity of carvacrol on human cancer cells. *International Journal of Pharmaceutical Research, 10*, 174-180.

Andersen, A. (2006). Final report on the safety assessment of sodium p-chloro-m-cresol, p-chloro-m-cresol, chlorothymol, mixed cresols, m-cresol, o-cresol, p-cresol, isopropyl cresols, thymol, o-cymen-5-ol, and carvacrol. *International Journal of Toxicology, 25*, 29-127.

Aravindaram, K., and Yang, N. S. (2010). Anti-inflammatory plant natural products for cancer therapy. *Planta Medica, 76*(11), 1103-1117.

Arigesavan, K., and Sudhandiran, G. (2015). Carvacrol exhibits anti-oxidant and anti-inflammatory effects against 1, 2-dimethyl hydrazine plus dextran sodium sulfate induced inflammation associated carcinogenicity in the colon of Fischer 344 rats. *Biochemical and Biophysical Research Communications, 461*(2), 314-320.

Aristatile, B., Al-Numair, K. S., Veeramani, C., and Pugalendi, K. V. (2009). Effect of carvacrol on hepatic marker enzymes and antioxidant status in d-galactosamine-induced hepatotoxicity in rats. *Fundamental and Clinical Pharmacology, 23*(6), 757-765.

Arunasree, K. (2010). Anti-proliferative effects of carvacrol on a human metastatic breast cancer cell line, MDA-MB 231. *Phytomedicine, 17*(8-9), 581-588.

Ashrafudoulla, M., Rahaman Mizan, M. F., Park, S. H., and Ha, S.-D. (2021). Antibiofilm activity of carvacrol against *Listeria monocytogenes* and *Pseudomonas aeruginosa* biofilm on MBEC™ biofilm device and polypropylene surface. *LWT, 147*, 111575. doi:https://doi.org/10.1016/j.lwt.2021.111575.

Aydın, S., Başaran, A. A., and Başaran, N. (2005a). The effects of thyme volatiles on the induction of DNA damage by the heterocyclic amine IQ and mitomycin C. *Mutation Research/Genetic Toxicology and Environmental Mutagenesis, 581*(1-2), 43-53.

Aydın, S., Başaran, A. A., and Başaran, N. (2005b). Modulating effects of thyme and its major ingredients on oxidative DNA damage in human lymphocytes. *Journal of Agricultural and Food Chemistry, 53*(4), 1299-1305.

Aziz, Z. A., Ahmad, A., Setapar, S. H. M., Karakucuk, A., Azim, M. M., Lokhat, D., . . . Ashraf, G. M. (2018). Essential oils: extraction techniques, pharmaceutical and therapeutic potential-a review. *Current Drug Metabolism, 19*(13), 1100-1110.

Azizi, Z., Salimi, M., Amanzadeh, A., Majelssi, N., and Naghdi, N. (2020). Carvacrol and thymol attenuate cytotoxicity induced by amyloid β25-35 via activating protein kinase C and inhibiting oxidative stress in PC12 cells. *Iranian Biomedical Journal, 24*(4), 243.

Bagci, Y., Kan, Y., Dogu, S., and Çelik, S. A. (2017). The essential oil compositions of *Origanum majorana* L. cultivated in Konya and collected from Mersin-Turkey. *Indian Journal of Pharmaceutical Education and Research, 51*, S463-S469.

Bakır, M., Geyikoglu, F., Colak, S., Turkez, H., Bakır, T. O., and Hosseinigouzdagani, M. (2016). The carvacrol ameliorates acute pancreatitis-induced liver injury via antioxidant response. *Cytotechnology, 68*(4), 1131-1146.

Baptista-Silva, S., Borges, S., Ramos, O. L., Pintado, M., and Sarmento, B. (2020). The progress of essential oils as potential therapeutic agents: A review. *Journal of Essential Oil Research, 32*(4), 279-295.

Bassanetti, I., Carcelli, M., Buschini, A., Montalbano, S., Leonardi, G., Pelagatti, P., . . . Rogolino, D. (2017). Investigation of antibacterial activity of new classes of essential oils derivatives. *Food Control, 73*, 606-612.

Baydar, H., Sağdiç, O., Özkan, G., and Karadoğan, T. (2004). Antibacterial activity and composition of essential oils from *Origanum, Thymbra* and

Satureja species with commercial importance in Turkey. *Food Control, 15*(3), 169-172.

Ben Arfa, A., Combes, S., Preziosi-Belloy, L., Gontard, N., and Chalier, P. (2006). Antimicrobial activity of carvacrol related to its chemical structure. *Letters in Applied Microbiology, 43*(2), 149-154.

Bernal-Mercado, A. T., Gutierrez-Pacheco, M. M., Encinas-Basurto, D., Mata-Haro, V., Lopez-Zavala, A. A., Islas-Osuna, M. A., . . . Ayala-Zavala, J. F. (2020). Synergistic mode of action of catechin, vanillic and protocatechuic acids to inhibit the adhesion of uropathogenic *Escherichia coli* on silicone surfaces. *Journal of Applied Microbiology, 128*(2), 387-400. doi:https://doi.org/10.1111/jam.14472.

Bhakkiyalakshmi, E., Suganya, N., Sireesh, D., Krishnamurthi, K., Devi, S. S., Rajaguru, P., and Ramkumar, K. M. (2016). Carvacrol induces mitochondria-mediated apoptosis in HL-60 promyelocytic and Jurkat T lymphoma cells. *European Journal of Pharmacology, 772*, 92-98.

Blowman, K., Magalhães, M., Lemos, M., Cabral, C., and Pires, I. (2018). Anticancer properties of essential oils and other natural products. *Evidence-Based Complementary and Alternative Medicine, 2018*.

Botelho, M., Nogueira, N., Bastos, G., Fonseca, S., Lemos, T., Matos, F., . . . Brito, G. (2007). Antimicrobial activity of the essential oil from *Lippia sidoides*, carvacrol and thymol against oral pathogens. *Brazilian Journal of Medical and Biological Research, 40*, 349-356.

Bouhtit, F., Najar, M., Moussa Agha, D., Melki, R., Najimi, M., Sadki, K., . . . Hamal, A. (2021). New anti-leukemic effect of carvacrol and thymol combination through synergistic induction of different cell death pathways. *Molecules, 26*(2), 410.

Bound, D. J., Murthy, P. S., Negi, P., and Srinivas, P. (2020). Evaluation of anti-quorum sensing and antimutagenic activity of 2, 3-unsaturated and 2, 3-dideoxyglucosides of terpene phenols and alcohols. *LWT, 122*, 108987.

Buchbauer, G., and Bohusch, R. (2016). Biological activities of essential oils an update. In K. H. Can Baser & G. Buchbauer (Eds.), *Handbook of Essential Oils: Science, Technology, and Applications* (pp. 281–307). Boca Raton, USA: CRC Press.

Busatta, C., Barbosa, J., Cardoso, R. I., Paroul, N., Rodrigues, M., Oliveira, D. d., . . . Cansian, R. L. (2017). Chemical profiles of essential oils of marjoram (*Origanum majorana*) and oregano (*Origanum vulgare*) obtained by hydrodistillation and supercritical CO_2. *Journal of Essential Oil Research, 29*(5), 367-374.

Cacciatore, I., Di Giulio, M., Fornasari, E., Di Stefano, A., Cerasa, L. S., Marinelli, L., . . . Robuffo, I. (2015). Carvacrol codrugs: a new approach in the antimicrobial plan. *PLoS One, 10*(4), e0120937.

Can Baser, K. (2008). Biological and pharmacological activities of carvacrol and carvacrol bearing essential oils. *Current Pharmaceutical Design, 14*(29), 3106-3119.

Chami, N., Bennis, S., Chami, F., Aboussekhra, A., and Remmal, A. (2005). Study of anticandidal activity of carvacrol and eugenol in vitro and in vivo. *Oral Microbiology and Immunology, 20*(2), 106-111. doi:https://doi.org/10.1111/j.1399-302X.2004.00202.x.

Ciandrini, E., Campana, R., Federici, S., Manti, A., Battistelli, M., Falcieri, E., . . . Baffone, W. (2014). In vitro activity of Carvacrol against titanium-adherent oral biofilms and planktonic cultures. *Clinical Oral Investigations, 18*(8), 2001-2013. doi:10.1007/s00784-013-1179-9.

Côté, H., Pichette, A., St-Gelais, A., and Legault, J. (2021). The biological activity of *Monarda didyma* L. essential oil and its effect as a diet supplement in mice and broiler chicken. *Molecules, 26*(11), 3368.

da Silva Lima, M., Quintans-Júnior, L. J., de Santana, W. A., Kaneto, C. M., Soares, M. B. P., and Villarreal, C. F. (2013). Anti-inflammatory effects of carvacrol: evidence for a key role of interleukin-10. *European Journal of Pharmacology, 699*(1-3), 112-117.

Damasceno, S. R., Oliveira, F. R. A., Carvalho, N. S., Brito, C. F., Silva, I. S., Sousa, F. B. M., . . . Freitas, R. M. (2014). Carvacryl acetate, a derivative of carvacrol, reduces nociceptive and inflammatory response in mice. *Life Sciences, 94*(1), 58-66.

de Carvalho, F. O., Silva, É. R., Gomes, I. A., Santana, H. S. R., do Nascimento Santos, D., de Oliveira Souza, G. P., . . . de Souza Araújo, A. A. (2020). Anti-inflammatory and antioxidant activity of carvacrol in the respiratory system: A systematic review and meta-analysis. *Phytotherapy Research, 34*(9), 2214-2229.

de Oliveira Pedrosa Rolim, M., Rodriguesde Almeida, A., Galdino da Rocha Pitta, M., Barreto de Melo Rêgo, M. J., Quintans-Junior, L. J., de Souza Siqueira Quintans, J., . . . Marques Duarte da Cruz, R. (2019). Design, synthesis and pharmacological evaluation of CVIB, a codrug of carvacrol and ibuprofen as a novel anti-inflammatory agent. *International Immunopharmacology, 76*, 105856.

De Vincenzi, M., Stammati, A., De Vincenzi, A., and Silano, M. (2004). Constituents of aromatic plants: carvacrol. *Fitoterapia, 75*(7-8), 801-804.

Deng, W., Lu, H., and Teng, J. (2013). Carvacrol attenuates diabetes-associated cognitive deficits in rats. *Journal of Molecular Neuroscience, 51*(3), 813-819.

Desai, M. A., Soni, K. A., Nannapaneni, R., Schilling, M. W., and Silva, J. L. (2012). Reduction of *Listeria monocytogenes* in Raw Catfish Fillets by Essential Oils and Phenolic Constituent Carvacrol. *Journal of Food Science, 77*(9), M516-M522. doi:https://doi.org/10.1111/j.1750-3841.2012.02859.x.

Dutra, T. V., Castro, J. C., Menezes, J. L., Ramos, T. R., do Prado, I. N., Junior, M. M., . . . de Abreu Filho, B. A. (2019). Bioactivity of oregano (*Origanum vulgare*) essential oil against *Alicyclobacillus* spp. *Industrial Crops and Products, 129*, 345-349.

Ebani, V. V., Nardoni, S., Bertelloni, F., Pistelli, L., and Mancianti, F. (2018). Antimicrobial activity of five essential oils against bacteria and fungi responsible for urinary tract infections. *Molecules, 23*(7), 1668.

El-Sayed, E. S. M., Mansour, A. M., and Abdul-Hameed, M. S. (2016). Thymol and carvacrol prevent doxorubicin-induced cardiotoxicity by abrogation of oxidative stress, inflammation, and apoptosis in rats. *Journal of Biochemical and Molecular Toxicology, 30*(1), 37-44.

Encyclopedia-Britannica, E. o. (2020). Inflammation. *Encyclopedia Britannica*. Retrieved from https://www.britannica.com/science/inflammation.

Engel, J. B., Heckler, C., Tondo, E. C., Daroit, D. J., and da Silva Malheiros, P. (2017). Antimicrobial activity of free and liposome-encapsulated thymol and carvacrol against *Salmonella* and *Staphylococcus aureus* adhered to stainless steel. *International Journal of Food Microbiology, 252*, 18-23.

Enioutina, E. Y., Salis, E. R., Job, K. M., Gubarev, M. I., Krepkova, L. V., and Sherwin, C. M. (2017). Herbal Medicines: challenges in the modern world. Part 5. status and current directions of complementary and alternative herbal medicine worldwide. *Expert Review of Clinical Pharmacology, 10*(3), 327-338.

Eusepi, P., Marinelli, L., Borrego-Sánchez, A., García-Villén, F., Rayhane, B. K., Cacciatore, I., . . . Di Stefano, A. (2020). Nano-delivery systems based on carvacrol prodrugs and fibrous clays. *Journal of Drug Delivery Science and Technology, 58*, 101815.

Ezz-Eldin, Y. M., Aboseif, A. A., and Khalaf, M. M. (2020). Potential anti-inflammatory and immunomodulatory effects of carvacrol against ovalbumin-induced asthma in rats. *Life Sciences, 242*, 117222.

Fan, K., Li, X., Cao, Y., Qi, H., Li, L., Zhang, Q., and Sun, H. (2015). Carvacrol inhibits proliferation and induces apoptosis in human colon cancer cells. *Anti-cancer Drugs, 26*(8), 813-823.

Fekrazad, R., Afzali, M., Pasban-Aliabadi, H., Esmaeili-Mahani, S., Aminizadeh, M., and Mostafavi, A. (2017). Cytotoxic effect of *Thymus caramanicus* Jalas on human oral epidermoid carcinoma KB cells. *Brazilian dental journal, 28*, 72-77.

Feng, X., and Jia, A. (2014). Protective effect of carvacrol on acute lung injury induced by lipopolysaccharide in mice. *Inflammation, 37*(4), 1091-1101.

Galehassadi, M., Rezaii, E., Najavand, S., Mahkam, M., and Mohammadzadeh, G. (2014). Isolation of carvacol from *Origanum vulgare*, synthesis of some organosilicon derivatives, and investigating of its antioxidant, antibacterial activities. *Standard Scientific Research and Essays, 2*, 438-450.

Gavaric, N., Mozina, S. S., Kladar, N., and Bozin, B. (2015). Chemical profile, antioxidant and antibacterial activity of thyme and oregano essential oils, thymol and carvacrol and their possible synergism. *Journal of Essential Oil Bearing Plants, 18*(4), 1013-1021.

Ghafari, O., Sharifi, A., Ahmadi, A., and Nayeri Fasaei, B. (2018). Antibacterial and anti-PmrA activity of plant essential oils against fluoroquinolone-resistant *Streptococcus pneumoniae* clinical isolates. *Letters in Applied Microbiology, 67*(6), 564-569. doi:https://doi.org/10.1111/lam.13050.

Ghorani, V., Alavinezhad, A., Rajabi, O., Mohammadpour, A. H., and Boskabady, M. H. (2021). Safety and tolerability of carvacrol in healthy subjects: a phase I clinical study. *Drug and Chemical Toxicology, 44*(2), 177-189.

Gilling, D. H., Kitajima, M., Torrey, J. R., and Bright, K. R. (2014). Antiviral efficacy and mechanisms of action of oregano essential oil and its primary component carvacrol against murine norovirus. *Journal of Applied Microbiology, 116*(5), 1149-1163. doi:https://doi.org/10.1111/jam.12453.

GLOBOCAN. (2020). *Global cancer observatory: cancer today*. Retrieved from https://www.uicc.org/news/globocan-2020-new-global-cancer-data.

González-Trujano, M. E., Hernández-Sánchez, L. Y., Muñoz Ocotero, V., Dorazco-González, A., Guevara Fefer, P., and Aguirre-Hernández, E. (2017). Pharmacological evaluation of the anxiolytic-like effects of *Lippia graveolens* and bioactive compounds. *Pharmaceutical Biology, 55*(1), 1569-1576.

Guimarães, A. G., Xavier, M. A., de Santana, M. T., Camargo, E. A., Santos, C. A., Brito, F. A., . . . Oliveira, R. C. (2012). Carvacrol attenuates mechanical hypernociception and inflammatory response. *Naunyn-Schmiedeberg's Archives of Pharmacology, 385*(3), 253-263.

Günes-Bayir, A., Kiziltan, H. S., Kocyigit, A., Güler, E. M., Karataş, E., and Toprak, A. (2017). Effects of natural phenolic compound carvacrol on the human gastric adenocarcinoma (AGS) cells *in vitro*. *Anti-cancer Drugs, 28*(5), 522-530.

Gupta, S. C., Kunnumakkara, A. B., Aggarwal, S., and Aggarwal, B. B. (2018). Inflammation, a double-edge sword for cancer and other age-related diseases. *Frontiers in Immunology, 9*, 2160.

Gutierrez-Pacheco, M., Gonzalez-Aguilar, G., Martinez-Tellez, M., Lizardi-Mendoza, J., Madera-Santana, T., Bernal-Mercado, A., . . . Ayala-Zavala, J. (2018). Carvacrol inhibits biofilm formation and production of extracellular polymeric substances of *Pectobacterium carotovorum* subsp. carotovorum. *Food Control, 89*, 210-218.

Güvenç, M., Cellat, M., Gökçek, İ., Yavaş, İ., and Yurdagül Özsoy, Ş. (2019). Effects of thymol and carvacrol on sperm quality and oxidant/antioxidant balance in rats. *Archives of Physiology and Biochemistry, 125*(5), 396-403.

Halliwell, B., and Gutteridge, J. M. (2015). *Free radicals in biology and medicine*: Oxford university press, USA.

Hashemi, S. M. B., Khaneghah, A. M., and Akbarirad, H. (2016). The effects of amplitudes ultrasound-assisted solvent extraction and pretreatment time on the yield and quality of *Pistacia khinjuk* hull oil. *Journal of oleo science, 65*(9), 733-738.

Hashemi, S. M. B., Khaneghah, A. M., Koubaa, M., Barba, F. J., Abedi, E., Niakousari, M., and Tavakoli, J. (2018). Extraction of essential oil from *Aloysia citriodora* Palau leaves using continuous and pulsed ultrasound: Kinetics, antioxidant activity and antimicrobial properties. *Process Biochemistry, 65*, 197-204.

Hassan, S. T., Berchová-Bímová, K., Šudomová, M., Malaník, M., Šmejkal, K., and Rengasamy, K. R. (2018). In vitro study of multi-therapeutic properties of *Thymus bovei* Benth. essential oil and its main component for promoting their use in clinical practice. *Journal of Clinical Medicine, 7*(9), 283.

Heidarian, E., and Keloushadi, M. (2019). Antiproliferative and anti-invasion effects of carvacrol on PC3 human prostate cancer cells through reducing pSTAT3, pAKT, and pERK1/2 signaling proteins. *International Journal of Preventive Medicine, 10*.

Hotta, M., Nakata, R., Katsukawa, M., Hori, K., Takahashi, S., and Inoue, H. (2010). Carvacrol, a component of thyme oil, activates PPARα and γ and suppresses COX-2 expression [S]. *Journal of Lipid Research, 51*(1), 132-139.

IARC. (2018). *Latest Global Cancer Data: Cancer Burden Rises to 18.1 million new cases and 9.6 million cancer deaths in 2018*. Retrieved from https://www.iarc.who.int/featured-news/latest-global-cancer-data-cancer-burden-rises-to-18-1-million-new-cases-and-9-6-million-cancer-deaths-in-2018/.

Ibáñez, M. D., López-Gresa, M. P., Lisón, P., Rodrigo, I., Bellés, J. M., González-Mas, M. C., and Blázquez, M. A. (2020). Essential oils as natural antimicrobial and antioxidant products in the agrifood indus. *Nereis. Revista Iberoamericana Interdisciplinar de Métodos, Modelización y Simulación* (12), 55-69.

Ipek, E., Zeytinoglu, H., Okay, S., Tuylu, B. A., Kurkcuoglu, M., and Baser, K. H. C. (2005). Genotoxicity and antigenotoxicity of *Origanum* oil and carvacrol evaluated by Ames *Salmonella*/microsomal test. *Food Chemistry, 93*(3), 551-556.

Jalalvand, M., Shahsavari, G., Sheikhian, A., Ganji, A., and Mosayebi, G. (2020). In vitro Anti-inflammatory Effects of Satureja Kkhuzestanica Essential Oil

Compared to Carvacrol. *International Journal of Basic Science in Medicine, 5*(2), 61-65.

Jayakumar, S., Madankumar, A., Asokkumar, S., Raghunandhakumar, S., Kamaraj, S., Divya, M. G. J., and Devaki, T. (2012). Potential preventive effect of carvacrol against diethylnitrosamine-induced hepatocellular carcinoma in rats. *Molecular and Cellular Biochemistry, 360*(1), 51-60.

Jornada, D. H., dos Santos Fernandes, G. F., Chiba, D. E., De Melo, T. R. F., Dos Santos, J. L., and Chung, M. C. (2016). The prodrug approach: A successful tool for improving drug solubility. *Molecules, 21*(1), 42.

Jung, C. Y., Kim, S. Y., and Lee, C. (2018). Carvacrol targets AXL to inhibit cell proliferation and migration in non-small cell lung cancer cells. *Anticancer research, 38*(1), 279-286.

Kachur, K., and Suntres, Z. (2020). The antibacterial properties of phenolic isomers, carvacrol and thymol. *Critical Reviews in Food Science and Nutrition, 60*(18), 3042-3053.

Karadayi, M., Yildirim, V., and Güllüce, M. Antimicrobial activity and other biological properties of oregano essential oil and carvacrol. *Anatolian Journal of Biology, 1*(2), 52-68.

Karygianni, L., Al-Ahmad, A., Argyropoulou, A., Hellwig, E., Anderson, A. C., and Skaltsounis, A. L. (2016). Natural antimicrobials and oral microorganisms: A systematic review on herbal interventions for the eradication of multispecies oral biofilms. *Frontiers in Microbiology, 6*(1529). doi:10.3389/fmicb.2015.01529.

Kavoosi, G., and Rabiei, F. (2015). *Zataria multiflora*: chemical and biological diversity in the essential oil. *Journal of Essential Oil Research, 27*(5), 428-436.

Khadir, A., Sobeh, M., Gad, H. A., Benbelaid, F., Bendahou, M., Peixoto, H., . . . Wink, M. (2016). Chemical composition and biological activity of the essential oil from *Thymus lanceolatus*. *Zeitschrift für Naturforschung C, 71*(5-6), 155-163.

Khalaf, A. A. A., Elhady, M. A., Hassanen, E. I., Azouz, A. A., Ibrahim, M. A., Galal, M. K., . . . Azouz, R. A. (2021). Antioxidant role of carvacrol against hepatotoxicity and nephrotoxicity induced by propiconazole in rats. *Revista Brasileira de Farmacognosia, 31*(1), 67-74.

Khalili, G., Mazloomifar, A., Larijani, K., Tehrani, M. S., and Azar, P. A. (2018). Solvent-free microwave extraction of essential oils from *Thymus vulgaris* L. and *Melissa officinalis* L. *Industrial Crops and Products, 119*, 214-217.

Khan, F., Khan, I., Farooqui, A., and Ansari, I. A. (2017). Carvacrol induces reactive oxygen species (ROS)-mediated apoptosis along with cell cycle arrest at G0/G1 in human prostate cancer cells. *Nutrition and Cancer, 69*(7), 1075-1087.

Khan, I., Bahuguna, A., Kumar, P., Bajpai, V. K., and Kang, S. C. (2017). Antimicrobial Potential of Carvacrol against Uropathogenic *Escherichia coli* via Membrane Disruption, Depolarization, and Reactive Oxygen Species Generation. *Frontiers in Microbiology, 8*(2421). doi:10.3389/fmicb.2017.02421.

Khan, I., Bahuguna, A., Kumar, P., Bajpai, V. K., and Kang, S. C. (2018). *In vitro* and *in vivo* antitumor potential of carvacrol nanoemulsion against human lung adenocarcinoma A549 cells via mitochondrial mediated apoptosis. *Scientific Reports, 8*(1), 1-15.

Khan, I., Bahuguna, A., Shukla, S., Aziz, F., Chauhan, A. K., Ansari, M. B., . . . Kang, S. C. (2020). Antimicrobial potential of the food-grade additive carvacrol against uropathogenic *E. coli* based on membrane depolarization, reactive oxygen species generation, and molecular docking analysis. *Microbial Pathogenesis, 142*, 104046. doi:https://doi.org/10.1016/j.micpath.2020.104046.

Khan, S. T., Khan, M., Ahmad, J., Wahab, R., Abd-Elkader, O. H., Musarrat, J., . . . Al-Kedhairy, A. A. (2017). Thymol and carvacrol induce autolysis, stress, growth inhibition and reduce the biofilm formation by *Streptococcus mutans*. *AMB Express, 7*(1), 49. doi:10.1186/s13568-017-0344-y.

Khazdair, M., Alavinezhad, A., and Boskabady, M. (2018). Carvacrol ameliorates haematological parameters, oxidant/antioxidant biomarkers and pulmonary function tests in patients with sulphur mustard-induced lung disorders: a randomized double-blind clinical trial. *Journal of Clinical Pharmacy and Therapeutics, 43*(5), 664-674.

Khazdair, M. R., and Boskabady, M. H. (2019). A double-blind, randomized, placebo-controlled clinical trial on the effect of carvacrol on serum cytokine levels and pulmonary function tests in sulfur mustard induced lung injury. *Cytokine, 113*, 311-318.

Kılıç, Y., Geyikoglu, F., Çolak, S., Turkez, H., Bakır, M., and Hsseinigouzdagani, M. (2016). Carvacrol modulates oxidative stress and decreases cell injury in pancreas of rats with acute pancreatitis. *Cytotechnology, 68*(4), 1243-1256.

Knez Hrnčič, M., Cör, D., Simonovska, J., Knez, Ž., Kavrakovski, Z., and Rafajlovska, V. (2020). Extraction techniques and analytical methods for characterization of active compounds in *Origanum* species. *Molecules, 25*(20), 4735.

Koraichi Saad, I., Hassan, L., Ghizlane, Z., Hind, M., and Adnane, R. (2011). Carvacrol and thymol components inhibiting *Pseudomonas aeruginosa* adherence and biofilm formation. *African Journal of Microbiology Research, 5*(20), 3229-3232.

Krisilia, V., Deli, G., Koutsaviti, A., and Tzakou, O. (2021). *Thymbra* L. and *Satureja* L. essential oils as rich sources of carvacrol, a food additive with health-promoting effects. *American Journal of Essential Oils Natural Products, 9*(1), 12-23.

Kulkarni, S. A., Nagarajan, S. K., Ramesh, V., Palaniyandi, V., Selvam, S. P., and Madhavan, T. (2020). Computational evaluation of major components from plant essential oils as potent inhibitors of SARS-CoV-2 spike protein. *Journal of Molecular Structure, 1221*, 128823. doi:https://doi.org/10.1016/j.molstruc.2020.128823.

Kumar, A., Choudhir, G., Shukla, S. K., Sharma, M., Tyagi, P., Bhushan, A., and Rathore, M. (2020). Identification of phytochemical inhibitors against main protease of COVID-19 using molecular modeling approaches. *Journal of Biomolecular Structure and Dynamics*, 1-11.

Landa, P., Kokoska, L., Pribylova, M., Vanek, T., and Marsik, P. (2009). In vitro anti-inflammatory activity of carvacrol: Inhibitory effect on COX-2 catalyzed prostaglandin E 2 biosynthesis. *Archives of Pharmacal Research, 32*(1), 75-78.

Landry, K. S., Chang, Y., McClements, D. J., and McLandsborough, L. (2014). Effectiveness of a novel spontaneous carvacrol nanoemulsion against *Salmonella enterica* Enteritidis and *Escherichia coli* O157:H7 on contaminated mung bean and alfalfa seeds. *International Journal of Food Microbiology, 187*, 15-21. doi:https://doi.org/10.1016/j.ijfoodmicro.2014.06.030.

Lee, J. H., Kim, Y. G., and Lee, J. (2017). Carvacrol-rich oregano oil and thymol-rich thyme red oil inhibit biofilm formation and the virulence of uropathogenic *Escherichia coli*. *Journal of Applied Microbiology, 123*(6), 1420-1428. doi:https://doi.org/10.1111/jam.13602.

Lim, W., Ham, J., Bazer, F. W., and Song, G. (2019). Carvacrol induces mitochondria-mediated apoptosis via disruption of calcium homeostasis in human choriocarcinoma cells. *Journal of Cellular Physiology, 234*(2), 1803-1815.

Liu, F., Jin, P., Sun, Z., Du, L., Wang, D., Zhao, T., and Doyle, M. P. (2021). Carvacrol oil inhibits biofilm formation and exopolysaccharide production of *Enterobacter cloacae*. *Food Control, 119*, 107473. doi:https://doi.org/10.1016/j.foodcont.2020.107473.

Lobo, V., Patil, A., Phatak, A., and Chandra, N. (2010). Free radicals, antioxidants and functional foods: Impact on human health. *Pharmacognosy Reviews, 4*(8), 118.

Magi, G., Marini, E., and Facinelli, B. (2015). Antimicrobial activity of essential oils and carvacrol, and synergy of carvacrol and erythromycin, against clinical, erythromycin-resistant Group A *Streptococci*. *Frontiers in Microbiology, 6*, 165.

Mahmoodi, M., Amiri, H., Ayoobi, F., Rahmani, M., Taghipour, Z., Ghavamabadi, R. T., . . . Sankian, M. (2019). Carvacrol ameliorates experimental autoimmune encephalomyelitis through modulating pro-and anti-inflammatory cytokines. *Life Sciences, 219*, 257-263.

Marchese, A., Arciola, C. R., Coppo, E., Barbieri, R., Barreca, D., Chebaibi, S., . . . Daglia, M. (2018). The natural plant compound carvacrol as an antimicrobial and anti-biofilm agent: mechanisms, synergies and bio-inspired anti-infective materials. *Biofouling, 34*(6), 630-656.

Mari, A., Mani, G., Nagabhishek, S. N., Balaraman, G., Subramanian, N., Mirza, F. B., . . . Thiruvengadam, D. (2020). Carvacrol promotes cell cycle arrest and apoptosis through PI3K/AKT signaling pathway in MCF-7 breast cancer cells. *Chinese Journal of Integrative Medicine*, 1-8.

Marinelli, L., Di Stefano, A., and Cacciatore, I. (2018). Carvacrol and its derivatives as antibacterial agents. *Phytochemistry Reviews, 17*(4), 903-921.

Marinelli, L., Fornasari, E., Eusepi, P., Ciulla, M., Genovese, S., Epifano, F., . . . Mingoia, M. (2019). Carvacrol prodrugs as novel antimicrobial agents. *European Journal of Medicinal Chemistry, 178*, 515-529.

Maritim, A., Sanders, a., and Watkins Iii, J. (2003). Diabetes, oxidative stress, and antioxidants: a review. *Journal of Biochemical and Molecular Toxicology, 17*(1), 24-38.

Maru, G. B., Hudlikar, R. R., Kumar, G., Gandhi, K., and Mahimkar, M. B. (2016). Understanding the molecular mechanisms of cancer prevention by dietary phytochemicals: From experimental models to clinical trials. *World journal of biological chemistry, 7*(1), 88-99. doi:10.4331/wjbc.v7.i1.88.

Mastelic, J., Jerkovic, I., Blažević, I., Poljak-Blaži, M., Borović, S., Ivančić-Baće, I., . . . Vikić-Topić, D. (2008). Comparative study on the antioxidant and biological activities of carvacrol, thymol, and eugenol derivatives. *Journal of Agricultural and Food Chemistry, 56*(11), 3989-3996.

Mehrjerdi, F. Z., Niknazar, S., Yadegari, M., Akbari, F. A., Pirmoradi, Z., and Khaksari, M. (2020). Carvacrol reduces hippocampal cell death and improves learning and memory deficits following lead-induced neurotoxicity via antioxidant activity. *Naunyn-Schmiedeberg's Archives of Pharmacology, 393*(7), 1229-1237.

Mezzoug, N., Elhadri, A., Dallouh, A., Amkiss, S., Skali, N., Abrini, J., . . . El Jaziri, M. (2007). Investigation of the mutagenic and antimutagenic effects of *Origanum compactum* essential oil and some of its constituents. *Mutation Research/Genetic Toxicology and Environmental Mutagenesis, 629*(2), 100-110.

Mihajilov-Krstev, T., Radnović, D., Kitić, D., Zlatković, B., Ristić, M., and Branković, S. (2009). Chemical composition and antimicrobial activity of *Satureja hortensis* L. essential oil. *Open Life Sciences, 4*(3), 411-416.

Mir, M., Permana, A. D., Ahmed, N., Khan, G. M., ur Rehman, A., and Donnelly, R. F. (2020). Enhancement in site-specific delivery of carvacrol for potential treatment of infected wounds using infection responsive nanoparticles loaded into dissolving microneedles: A proof of concept study. *European Journal of Pharmaceutics and Biopharmaceutics, 147*, 57-68.

Mohammadzamani, Z., Khorshidi, A., Khaledi, A., Shakerimoghaddam, A., Moosavi, G. A., and Piroozmand, A. (2020). Inhibitory effects of Cinnamaldehyde, Carvacrol, and honey on the expression of *exoS* and *ampC* genes in multidrug-resistant *Pseudomonas aeruginosa* isolated from burn wound infections. *Microbial Pathogenesis, 140*, 103946.

Montagu, A., Joly-Guillou, M.-L., Rossines, E., Cayon, J., Kempf, M., and Saulnier, P. (2016). Stress Conditions Induced by Carvacrol and Cinnamaldehyde on *Acinetobacter baumannii*. *Frontiers in Microbiology, 7*(1133). doi:10.3389/fmicb.2016.01133.

Mooyottu, S., Flock, G., Upadhyay, A., Upadhyaya, I., Maas, K., and Venkitanarayanan, K. (2017). Protective effect of carvacrol against gut dysbiosis and *Clostridium difficile* associated disease in a mouse model. *Frontiers in Microbiology, 8*(625). doi:10.3389/fmicb.2017.00625.

Morshedloo, M. R., Salami, S. A., Nazeri, V., Maggi, F., and Craker, L. (2018). Essential oil profile of oregano *(Origanum vulgare* L.) populations grown under similar soil and climate conditions. *Industrial Crops Products, 119*, 183-190.

Mortazavi, A., Kargar, H. M. P., Beheshti, F., Anaeigoudari, A., Vaezi, G., and Hosseini, M. (2021). The effects of carvacrol on oxidative stress, inflammation, and liver function indicators in a systemic inflammation model induced by lipopolysaccharide in rats. *International Journal for Vitamin and Nutrition Research*.

Mosaddad, S. A., Tahmasebi, E., Yazdanian, A., Rezvani, M. B., Seifalian, A., Yazdanian, M., and Tebyanian, H. (2019). Oral microbial biofilms: an update. *European Journal of Clinical Microbiology & Infectious Diseases, 38*(11), 2005-2019. doi:10.1007/s10096-019-03641-9.

Mousavi, S., Schmidt, A.-M., Escher, U., Kittler, S., Kehrenberg, C., Thunhorst, E., . . . Heimesaat, M. M. (2020). Carvacrol ameliorates acute campylobacteriosis in a clinical murine infection model. *Gut Pathogens, 12*(1), 2. doi:10.1186/s13099-019-0343-4.

Murdoch, D. R., and Howie, S. R. (2018). The global burden of lower respiratory infections: making progress, but we need to do better. *The Lancet Infectious Diseases, 18*(11), 1162-1163.

Mutlu-Ingok, A., Catalkaya, G., Capanoglu, E., and Karbancioglu-Guler, F. (2021). Antioxidant and antimicrobial activities of fennel, ginger, oregano and thyme essential oils. *Food Frontiers.*

Negrut, N., Khan, S. A., Bungau, S., Zaha, D. C., Aron, C., Bratu, O., . . . Ionita-Radu, F. (2020). Diagnostic challenges in gastrointestinal infections. *Romanian Journal of Military Medicine, 123*, 83-90.

Niaz, T., Imran, M., and Mackie, A. (2021). Improving carvacrol bioaccessibility using core–shell carrier-systems under simulated gastrointestinal digestion. *Food Chemistry, 353*, 129505.

Nicolle, L. E. (2014). Catheter associated urinary tract infections. *Antimicrobial Resistance and Infection Control, 3*(1), 23. doi:10.1186/2047-2994-3-23.

Nobrega, R. d. O., Teixeira, A. P. d. C., Oliveira, W. A. d., Lima, E. d. O., and Lima, I. O. (2016). Investigation of the antifungal activity of carvacrol against strains of *Cryptococcus neoformans*. *Pharmaceutical Biology, 54*(11), 2591-2596.

Nostro, A., and Papalia, T. (2012). Antimicrobial activity of carvacrol: current progress and future prospectives. *Recent Patents on Anti-infective Drug Discovery, 7*(1), 28-35.

Novak, J., Lukas, B., and Franz, C. (2010). Temperature influences thymol and carvacrol differentially in *Origanum* spp. (Lamiaceae). *Journal of Essential Oil Research, 22*(5), 412-415.

Obaidat, R. M., Bader, A., Al-Rajab, W., ABU SHEIKHA, G., and Obaidat, A. A. (2011). Preparation of mucoadhesive oral patches containing tetracycline hydrochloride and carvacrol for treatment of local mouth bacterial infections and candidiasis. *Scientia Pharmaceutica, 79*(1), 197-212.

Oezbek, T., Guelluece, M., Şahin, F., Oezkan, H., Sevsay, S., and Bariş, Ö. (2008). Investigation of the antimutagenic potentials of the methanol extract of *Origanum vulgare* L. subsp. *vulgare* in the eastern anatolia region of Turkey. *Turkish Journal of Biology, 32*(4), 271-276.

Önal, İ. Ö., Erden, İ. A., Akinci, S. B., Erden, G., Karabulut, İ., Zeybek, N. D., . . . Balkancioğlu, Z. D. (2011). The effect of carvacrol in oleic acid induced acute lung injury in rats [Sıçamlarda karvakrol'un oleik asitle oluşturulan akut akciğeri hasarına etkisi]. *Anestezi Dergisi, 19*(2), 90-98.

Orhan, I. E., Ozcelik, B., Kan, Y., and Kartal, M. (2011). Inhibitory Effects of Various Essential Oils and Individual Components against Extended-Spectrum Beta-Lactamase (ESBL) Produced by *Klebsiella pneumoniae* and Their Chemical Compositions. *Journal of Food Science, 76*(8), M538-M546. doi:https://doi.org/10.1111/j.1750-3841.2011.02363.x.

Özkan, O. E., Güney, K., Gür, M., Pattabanoğlu, E. S., Babat, E., and Khalifa, M. M. (2017). Essential oil of oregano and savory: Chemical composition and antimicrobial activity. *Indian Journal of Pharmaceutical Education and Research, 51*(3), S205-S208.

Peralta-Pérez, M. d. R., and Volke-Sepúlveda, T. (2012). La defensa antioxidante en las plantas: una herramienta clave para la fitorremediación. *Revista Mexicana de Ingeniería Química, 11*(1), 75-88.

Pilau, M. R., Alves, S. H., Weiblen, R., Arenhart, S., Cueto, A. P., and Lovato, L. T. (2011). Antiviral activity of the *Lippia graveolens* (Mexican oregano) essential oil and its main compound carvacrol against human and animal viruses. *Brazilian Journal of Microbiology, 42*, 1616-1624.

Pisoschi, A. M., and Pop, A. (2015). The role of antioxidants in the chemistry of oxidative stress: A review. *European Journal of Medicinal Chemistry, 97*, 55-74.

Pohanka, M. (2014). Alzheimer's disease and oxidative stress: a review. *Current Medicinal Chemistry, 21*(3), 356-364.

Potterat, O., and Hamburger, M. (2008). Drug discovery and development with plant-derived compounds. *Natural Compounds as Drugs Volume I*, 45-118.

Pourhosseini, S. H., Ahadi, H., Aliahmadi, A., and Mirjalili, M. H. (2020). Chemical composition and antibacterial activity of the carvacrol-rich essential oils of *Zataria multiflora* Boiss. (Lamiaceae) from southern natural habitats of Iran. *Journal of Essential Oil Bearing Plants, 23*(4), 779-787.

Radünz, M., da Trindade, M. L. M., Camargo, T. M., Radünz, A. L., Borges, C. D., Gandra, E. A., and Helbig, E. (2019). Antimicrobial and antioxidant activity of unencapsulated and encapsulated clove (*Syzygium aromaticum* L.) essential oil. *Food Chemistry, 276*, 180-186.

Ragno, R., Papa, R., Patsilinakos, A., Vrenna, G., Garzoli, S., Tuccio, V., . . . Artini, M. (2020). Essential oils against bacterial isolates from cystic fibrosis patients by means of antimicrobial and unsupervised machine learning approaches. *Scientific Reports, 10*(1), 1-11.

Raut, J. S., and Karuppayil, S. M. (2014). A status review on the medicinal properties of essential oils. *Industrial Crops and Products, 62*, 250-264. doi:https://doi.org/10.1016/j.indcrop.2014.05.055.

Rea, I. M., Gibson, D. S., McGilligan, V., McNerlan, S. E., Alexander, H. D., and Ross, O. A. (2018). Age and age-related diseases: role of inflammation triggers and cytokines. *Frontiers in Immunology, 9*, 586.

Reuter, S., Gupta, S. C., Chaturvedi, M. M., and Aggarwal, B. B. (2010). Oxidative stress, inflammation, and cancer: how are they linked? *Free Radical Biology and Medicine, 49*(11), 1603-1616.

Rivas, L., McDonnell, M. J., Burgess, C. M., O'Brien, M., Navarro-Villa, A., Fanning, S., and Duffy, G. (2010). Inhibition of verocytotoxigenic *Escherichia coli* in model broth and rumen systems by carvacrol and thymol.

International Journal of Food Microbiology, 139(1), 70-78. doi:https://doi.org/10.1016/j.ijfoodmicro.2010.01.029.

Rock, K. L., and Kono, H. (2008). The inflammatory response to cell death. *Annual Review of Pathology: Mechanisms of Disease, 3*, 99-126.

Rodrigues, M. R. A., Krause, L. C., Caramão, E. B., dos Santos, J. G., Dariva, C., and Vladimir de Oliveira, J. (2004). Chemical composition and extraction yield of the extract of *Origanum vulgare* obtained from sub-and supercritical CO_2. *Journal of Agricultural and Food Chemistry, 52*(10), 3042-3047.

Rodriguez-Garcia, I., Cruz-Valenzuela, M. R., Silva-Espinoza, B. A., Gonzalez-Aguilar, G. A., Moctezuma, E., Gutierrez-Pacheco, M. M., . . . Ayala-Zavala, J. F. (2016). Oregano (*Lippia graveolens*) essential oil added within pectin edible coatings prevents fungal decay and increases the antioxidant capacity of treated tomatoes. *Journal of the Science of Food and Agriculture, 96*(11), 3772-3778.

Rota, M. C., Herrera, A., Martínez, R. M., Sotomayor, J. A., and Jordán, M. J. (2008). Antimicrobial activity and chemical composition of *Thymus vulgaris*, *Thymus zygis* and *Thymus hyemalis* essential oils. *Food Control, 19*(7), 681-687.

Sahraoui, N., Hazzit, M., and Boutekedjiret, C. (2017). Effects of microwave heating on the antioxidant and insecticidal activities of essential oil of *Origanum glandulosum* Desf. obtained by microwave steam distillation. *Journal of Essential Oil Research, 29*(5), 420-429.

Samarghandian, S., Farkhondeh, T., Samini, F., and Borji, A. (2016). Protective effects of carvacrol against oxidative stress induced by chronic stress in rat's brain, liver, and kidney. *Biochemistry Research International, 2016*.

Shanaida, M., Jasicka-Misiak, I., Bialon, M., Korablova, O., and Wieczorek, P. P. (2021). Chromatographic profiles of the main secondary metabolites in the *Monarda fistulosa* L. aerial part. *Research Journal of Pharmacy and Technology, 14*(4), 2179-2184.

Sharifi-Rad, M., Varoni, E. M., Iriti, M., Martorell, M., Setzer, W. N., del Mar Contreras, M., . . . Tajbakhsh, M. (2018). Carvacrol and human health: A comprehensive review. *Phytotherapy Research, 32*(9), 1675-1687.

Sherwood, E. R., and Toliver-Kinsky, T. (2004). Mechanisms of the inflammatory response. *Best Practice and Research Clinical Anaesthesiology, 18*(3), 385-405.

Shinde, P., Agraval, H., Srivastav, A. K., Yadav, U. C., and Kumar, U. (2020). Physico-chemical characterization of carvacrol loaded zein nanoparticles for enhanced anticancer activity and investigation of molecular interactions between them by molecular docking. *International Journal of Pharmaceutics, 588*, 119795.

Shoorei, H., Khaki, A., Khaki, A. A., Hemmati, A. A., Moghimian, M., and Shokoohi, M. (2019). The ameliorative effect of carvacrol on oxidative stress and germ cell apoptosis in testicular tissue of adult diabetic rats. *Biomedicine and Pharmacotherapy, 111*, 568-578.

Shrestha, S., Wagle, B. R., Upadhyay, A., Arsi, K., Donoghue, D. J., and Donoghue, A. M. (2019). Carvacrol antimicrobial wash treatments reduce *Campylobacter jejuni* and aerobic bacteria on broiler chicken skin. *Poultry Science, 98*(9), 4073-4083. doi:https://doi.org/10.3382/ps/pez198.

Simirgiotis, M. J., Burton, D., Parra, F., López, J., Muñoz, P., Escobar, H., and Parra, C. (2020). Antioxidant and antibacterial capacities of *Origanum vulgare* l. Essential oil from the arid andean region of chile and its chemical characterization by GC-MS. *Metabolites, 10*(10), 414.

Słoczyńska, K., Powroźnik, B., Pękala, E., and Waszkielewicz, A. M. (2014). Antimutagenic compounds and their possible mechanisms of action. *Journal of applied genetics, 55*(2), 273-285.

Soni, K. A., Oladunjoye, A., Nannapaneni, R., Schilling, M. W., Silva, J. L., Mikel, B., and Bailey, R. H. (2013). Inhibition and inactivation of *Salmonella* Typhimurium biofilms from polystyrene and stainless steel surfaces by essential oils and phenolic constituent carvacrol. *Journal of Food Protection, 76*(2), 205-212.

Soto-Armenta, L., Sacramento-Rivero, J., Acereto-Escoffié, P., Peraza-González, E., Reyes-Sosa, C., and Rocha-Uribe, J. (2017). Extraction yield of essential oil from *Lippia graveolens* leaves by steam distillation at laboratory and pilot scales. *Journal of Essential Oil Bearing Plants, 20*(3), 610-621.

Souza, A. C. A., Abreu, F. F., Diniz, L. R., Grespan, R., DeSantana, J. M., Quintans-Júnior, L. J., . . . Teixeira, S. A. (2018). The inclusion complex of carvacrol and β-cyclodextrin reduces acute skeletal muscle inflammation and nociception in rats. *Pharmacological Reports, 70*(6), 1139-1145.

Stratakos, A. C., Sima, F., Ward, P., Linton, M., Kelly, C., Pinkerton, L., . . . Corcionivoschi, N. (2018). The in vitro effect of carvacrol, a food additive, on the pathogenicity of O157 and non-O157 Shiga-toxin producing *Escherichia coli*. *Food Control, 84*, 290-296. doi:https://doi.org/10.1016/j.foodcont.2017.08.014.

Suntres, Z. E., Coccimiglio, J., and Alipour, M. (2015). The bioactivity and toxicological actions of carvacrol. *Critical Reviews in Food Science and Nutrition, 55*(3), 304-318.

Swetha, T. K., Vikraman, A., Nithya, C., Hari Prasath, N., and Pandian, S. K. (2020). Synergistic antimicrobial combination of carvacrol and thymol impairs single and mixed-species biofilms of *Candida albicans* and *Staphylococcus epidermidis*. *Biofouling, 36*(10), 1256-1271.

Tapia-Rodriguez, M. R., Bernal-Mercado, A. T., Gutierrez-Pacheco, M. M., Vazquez-Armenta, F. J., Hernandez-Mendoza, A., Gonzalez-Aguilar, G. A., . . . signaling. (2019). Virulence of *Pseudomonas aeruginosa* exposed to carvacrol: Alterations of the Quorum sensing at enzymatic and gene levels. *Journal of Cell Communication, 13*(4), 531-537.

Tavakolpour, Y., Moosavi-Nasab, M., Niakousari, M., Haghighi-Manesh, S., Hashemi, S. M. B., and Mousavi Khaneghah, A. (2017). Comparison of four extraction methods for essential oil from *Thymus daenensis* Subsp. Lancifolius and chemical analysis of extracted essential oil. *Journal of Food Processing and Preservation, 41*(4), e13046.

Ternhag, A., Törner, A., Svensson, A., Ekdahl, K., and Giesecke, J. (2008). Short- and long-term effects of bacterial gastrointestinal infections. *Emerging infectious diseases, 14*(1), 143-148. doi:10.3201/eid1401.070524.

Tongnuanchan, P., and Benjakul, S. (2014). Essential oils: extraction, bioactivities, and their uses for food preservation. *Journal of Food Science, 79*(7), R1231-R1249.

Trindade, G. G., Thrivikraman, G., Menezes, P. P., França, C. M., Lima, B. S., Carvalho, Y. M., . . . Quintans-Júnior, L. (2019). Carvacrol/β-cyclodextrin inclusion complex inhibits cell proliferation and migration of prostate cancer cells. *Food and Chemical Toxicology, 125*, 198-209.

Ündeğer, Ü., Başaran, A., Degen, G., and Başaran, N. (2009). Antioxidant activities of major thyme ingredients and lack of (oxidative) DNA damage in V79 Chinese hamster lung fibroblast cells at low levels of carvacrol and thymol. *Food and Chemical Toxicology, 47*(8), 2037-2043.

Upadhyay, A., Arsi, K., Wagle, B. R., Upadhyaya, I., Shrestha, S., Donoghue, A. M., and Donoghue, D. J. (2017). Trans-cinnamaldehyde, carvacrol, and eugenol reduce *Campylobacter jejuni* colonization factors and expression of virulence genes in vitro. *Frontiers in Microbiology, 8*(713). doi:10.3389/fmicb.2017.00713.

Vamanu, E., Dinu, L. D., Luntraru, C. M., and Suciu, A. (2021). *In vitro* coliform resistance to bioactive compounds in urinary infection, assessed in a lab catheterization model. *Applied Sciences, 11*(9), 4315.

Wang, G. P. (2015). Defining functional signatures of dysbiosis in periodontitis progression. *Genome Medicine, 7*(1), 40. doi:10.1186/s13073-015-0165-z.

Wang, T.-H., Hsia, S.-M., Wu, C.-H., Ko, S.-Y., Chen, M. Y., Shih, Y.-H., . . . Wu, C.-Y. (2016). Evaluation of the Antibacterial Potential of Liquid and Vapor Phase Phenolic Essential Oil Compounds against Oral Microorganisms. *PLoS One, 11*(9), e0163147. doi:10.1371/journal.pone.0163147.

Wang, Z., Ding, L., Li, T., Zhou, X., Wang, L., Zhang, H., . . . Wang, H. (2006). Improved solvent-free microwave extraction of essential oil from dried *Cuminum cyminum* L. and *Zanthoxylum bungeanum* Maxim. *Journal of Chromatography A, 1102*(1-2), 11-17.

WHO, W. H. O. (2017). *Diarrhoeal disease.* Fact Sheet Retrieved from https://www.who.int/news-room/fact-sheets/detail/diarrhoeal-disease.

Wu, X., Patterson, S., and Hawk, E. (2011). Chemoprevention–history and general principles. *Best Practice and Research Clinical Gastroenterology, 25*(4-5), 445-459.

Yin, Q.-h., Yan, F.-x., Zu, X.-Y., Wu, Y.-h., Wu, X.-p., Liao, M.-c., . . . Zhuang, Y.-z. (2012). Anti-proliferative and pro-apoptotic effect of carvacrol on human hepatocellular carcinoma cell line HepG-2. *Cytotechnology, 64*(1), 43-51.

Zanini, S., Silva-Angulo, A., Rosenthal, A., Rodrigo, D., and Martínez, A. (2014). Effect of citral and carvacrol on the susceptibility of *Listeria monocytogenes* and *Listeria innocua* to antibiotics. *Letters in Applied Microbiology, 58*(5), 486-492.

Zgheib, R., El-Beyrouthy, M., Chaillou, S., Ouaini, N., Rutledge, D. N., Stien, D., . . . Iriti, M. (2019). Chemical variability of the essential oil of *Origanum ehrenbergii* boiss. From Lebanon, assessed by independent component analysis (ICA) and common component and specific weight analysis (CCSWA). *International Journal of Molecular Sciences, 20*(5), 1026.

Zhang, J., Onakpoya, I. J., Posadzki, P., and Eddouks, M. (2015). The safety of herbal medicine: from prejudice to evidence. *Evidence-Based Complementary and Alternative Medicine, 2015*, 316706. doi:10.1155/2015/316706.

Zhang, N., Wang, L., Deng, X., Liang, R., Su, M., He, C., . . . Jiang, S. (2020). Recent advances in the detection of respiratory virus infection in humans. *Journal of Medical Virology, 92*(4), 408-417. doi:https://doi.org/10.1002/jmv.25674.

Zumla, A., and Niederman, M. S. (2020). Editorial: The explosive epidemic outbreak of novel coronavirus disease 2019 (COVID-19) and the persistent threat of respiratory tract infectious diseases to global health security. *Current opinion in pulmonary medicine, 26*(3), 193-196. doi:10.1097/MCP.0000000000000676.

Chapter 2

Carvacrol as an Additional Barrier for Control of Pathogens during the Thermal Processing of Meat

Martin Valenzuela-Melendres[1], María González-Leyva[2] and Jorge I. López-Pino[1,*]

[1]Coordinación de Tecnología de Alimentos de Origen Animal,
Centro de Investigación en Alimentación y Desarrollo,
Hermosillo, Sonora, México
[2]Coordinación de Tecnología de Alimentos de Origen Vegetal,
Centro de Investigación en Alimentación y Desarrollo,
Hermosillo, Sonora, México

Abstract

The main guarantee against foodborne illness is exposure to heat to eliminate pathogens present in meat and meat products. Additional barriers, such as pH control, water activity, and the addition of antimicrobials such as carvacrol, are used to increase safety margins during heat treatments of food. Carvacrol is a monoterpene phenol produced by various aromatic plants, and it is used in low concentrations as a flavoring and seasoning ingredient for food and has significant antimicrobial activity against the principal pathogens. Carvacrol interacts with the meat matrix components, mainly fat, and has a significant

---

[*] Corresponding Author's E-mail: jrgisaac@gmail.com.

In: A Closer Look at Carvacrol
Editor: Zak A. Cunningham
ISBN: 978-1-68507-627-6
© 2022 Nova Science Publishers, Inc.

impact on the heat treatments of food. This compound affects the behavior of pathogens during the thermal process, making them more susceptible to the effect of temperature. However, since the direct use of carvacrol affects the product's sensory properties, it is necessary to use technologies such as atomization microencapsulation for its incorporation. The use of carvacrol as part of barrier technology in the thermal processing of meat products is a very promising strategy to limit and control the growth of pathogenic microorganisms.

Keywords: natural antimicrobial, meat safety, foodborne pathogens

Essential Oil as Natural Antimicrobial Agents in Meat Products

The growing demand for healthy and chemical-free foods has led to the investigation of new sources of natural preservatives. Plants and their derivatives, such as essential oils, are often used in folk medicine and have been recognized for years for their high efficacy as antimicrobials. In nature, essential oils play a critical role in the plants' defenses and can be obtained from different parts of plants such as leaves, roots, and flowers (Nazzaro et al., 2013). Essential oils contain a wide variety of secondary metabolites capable of inhibiting or retarding the growth of bacteria, yeasts, and molds. These combine the antioxidant and antimicrobial activity of their components against a variety of targets, particularly the membrane and cytoplasm, and in some cases, completely change the morphology of cells (Bouhdid et al., 2009).

Among the essential oils extracted from aromatic plants, oregano (*Origanum vulgare*), thyme (*Thymus vulgaris*), and rosemary (*Rosmarinus officinalis*) have been used in biomedical and industrial applications. The recent emergence of bacteria resistant to multiple antibiotics has stimulated research on the use of naturally occurring compounds as an alternative to traditional chemical compounds used to combat pathogenic microorganisms (Villa and Veiga-Crespo 2013). Essential oils and their components are naturally associated as antimicrobials due to their functions within the defense system in plants (Maldonado et al., 2015). Among the aromatic compounds that we can find in the essential oils of numerous plants are thymol, carvacrol, cinnamaldehyde, among others.

Carvacrol is a monoterpene phenolic compound produced by several aromatic plants, including *O. vulgare*, *O. dictamnus*, and *T. vulgaris*. These plants are used in low concentrations as a flavoring and seasoning ingredient for foods and exert significant antimicrobial activity against foodborne pathogens (Scandorieiro et al., 2016). There are numerous reports of the use of carvacrol as an antibacterial due to its action on cells, such as destabilization and rupture of the membrane (Gomes Neto et al., 2015); inhibition of reactions at the membrane level (Picone et al., 2013); damage to membrane proteins; decreased ionic strength of protons (de Souza 2016) or inhibition of antibiotic exit mechanisms from the cell (Miladi et al., 2016).

The use of essential oils as preservatives in meat and minimally processed products is part of the new trend of using green technologies and organic products, moving away from traditional chemical compounds. The supplementation of these compounds of natural origin to animal diet has shown favorable results due to the accumulation in the muscle of sufficient quantities to have an antimicrobial effect on the contaminating microbiota once the animal has been slaughtered. These essential oils have a long history as safe components for consumers, being accepted by regulatory bodies. In addition, more than 100 essential oils are considered safe for consumption that can be used to improve the microbiological quality of meat and meat products without affecting the consumers' health (Falleh et al., 2020).

The application of essential oils in the meat is made done in different ways; one of the most used is indirect, applying them as container components. For example, active packaging or pads at the bottom of them as support for essential oils has been used to prevent the proliferation of pathogens in fresh meat and some ready-to-eat meat products (Sharma et al., 2020). The application of emulsions composed of several essential oils provides better results than the separate components for inhibiting the growth of pathogenic microorganisms. For example, the mixtures of thymol, carvacrol, and linalool are effective against Gram-positive psychrophilic microorganisms such as *S. aureus* or *E. faecalis* and Gram-negative microorganisms such as *P. fluorescens*, *P. putida*, *C. jejuni*, or *S. enterica* (Agrimonti et al., 2019). The significant reduction in the microbial load reported in this work indicates that this method of indirect application of essential oils can be a viable alternative to improve the microbial quality of meat, especially at the level of retail sales where this strategy already it is widely used in other fresh foods.

In a study carried out in 2015 (Sarrazin et al., 2015), using oregano essential oil, against *B. cereus*, *B. subtili* and *S. typhimurium*, with a minimum inhibitory concentration of 0.62 µL/mL and 1.25 µL/ mL, were obtained inhibition halos greater than 25 mm (strong inhibition ≥ 20 mm) in both cases after 3 h direct exposure to the compound. On the other hand, in a study using two leafy vegetables: Swiss chard inoculated with *Escherichia coli*, *Listeria monocytogenes*, and *Salmonella enteritidis* and essential oils of oregano (*O. vulgare* L.) and rosemary (*R. officinalis* L.) as natural antimicrobials, the authors demonstrated that it is possible to reduce contamination levels by 3 log CFU/g with 10 min of exposure of the microorganism to the antimicrobial (de Medeiros Barbosa et al., 2016).

One of the microorganisms most regulated by the impact on the health of consumers is *E. coli*, this bacterium that is part of the native microbiota of pigs and cattle has been a serious health problem in the past. The application of essential oils of oregano, thyme, cinnamon, cloves, among others, have shown an effect on metabolism, stability of cell structures, and genes linked to the pathogenicity of *E. coli* strains. Some studies developed to evaluate the effect of the MIC of various essential oils on *E. coli* O157: H7 have demonstrated the efficacy of these to inhibit the synthesis of the toxin follow, from the regulation of several genes involved in the synthesis mechanisms and expression (Munekata et al., 2020). Heat treatment is the primary tool used to control these pathogens, and during the dried meat production process, the application of oregano essential oil is an additional barrier to increasing the safety of this product. The addition of oregano essential oil decreased the microbial load and increased sensory acceptability and demonstrated that it could be an alternative to chemical preservatives such as sodium nitrite in dried beef (Hernández et al., 2017).

Incorporating essential oils into marinating mixtures is another of the most used ways to apply these compounds. The work developed by Siroli et al. (2020) incorporates oregano, rosemary, and juniper oils into vacuum-packed pork marinating mix. The contaminating microbial load decreased significantly with the application of these compounds in the mixture, and contributed to the sensory attributes of the product. The extension of the shelf life of this type of preparation is highly valued in the meat industry because they can reduce and control the populations of pathogens such as *L. monocytogenes*, *S. aureus*, and *S. enteritidis*, which are some of the main pathogens that affect the microbiological quality of meat (Siroli et al., 2020)

Upper case antibiotic-resistant bacteria have been reported in food processing chains, like dairy products, meat, poultry, fish, and shellfish. The implementation of barrier technologies to stop and eliminate these microorganisms includes heat treatment, chemical agents, UV light, high pressures, and their combination with natural antimicrobials such as essential oils that have proven to be very effective in the control of numerous pathogens. Among the microorganisms that have been reported and that pose a high risk to the health of consumers are *B. cereus, C. perfringens, S. aureus, C. jejuni,*

S. entererica, *E. coli*, *V. cholerae*, and *V. parahemolyticus*. These pathogens reach meat through cross-contamination during gutting or poor production practices. Some of these microorganisms can withstand low storage temperatures and conditions in production areas where meat is handled, adding additional risk to the consumer.

The ease and speed with which antibiotic resistance genes spread from one microorganism to another pose a challenge for the meat industry. The BLES genes that generate the mechanisms of resistance to antibiotics generate evasive responses to traditional drugs, but they cannot deal with naturally occurring compounds. Thyme, oregano, or cinnamon essential oils cause a decrease in the expression of antibiotic resistance genes, and in addition, they cause irreversible damage to cell membranes that are incompatible with the life of these microorganisms. In the same way, they reduce the thermal resistance of pathogens when they are present in meat, making heat treatments more effective, and there is no effect on the nutritional quality of food caused by overcooking to eliminate pathogens (Friedman 2015).

Essential oils have shown high effectiveness for controlling the pathogen in meat products, and their effect is more noticeable on Gram-positive bacteria. By presenting the LPS layer on the outside of the membrane, Gram-negative bacteria cause a decrease in the permeability of the hydrophobic compounds that make up the oils, limiting their bactericidal effect. In some cases, the sensitivity to essential oils or their components may vary; for example, *Aeromonas hydrophila* has a high sensitivity to this type of compound, while bacteria of the genus *Pseudomonas* have a high resistance to the effect of essential oils. This result is not the same for all antimicrobials since clove, cinnamon, and oregano oils have the same effect on the two groups of bacteria (Aminzare et al., 2016).

The antimicrobial mechanism of essential oils obtained from oregano, rosemary, thyme, or cloves is varied. The main components of these essential oils, such as carvacrol, thymol, or eugenol, interact with membranes, efflux pumps, genetic material, among others, causing a potent bactericidal effect that has been used through its application in many meat products. Interactions between essential oils and the rest of the components of meat and meat products also influence the antimicrobial effect due to the complex relationships that arise during heat treatment and storage.

Antimicrobial Mechanism of Carvacrol

Different mechanisms of action have been proposed by which carvacrol affects the dynamic balance and homeostasis of foodborne microorganisms. The main impact is produced by alterations in the cytoplasmic membrane, where changes in permeability modify proton gradients. The ionic imbalance affects ATP synthesis and the electrical potential of the membrane. The effect on the membrane potential interferes with the transport of particles through it, causing coagulation of the cellular content by modifying the pH of the cytoplasm. In addition there are reports of membrane rupture and leakage of cell content, which significantly affects development and inhibits the spread of pathogens. It has also been shown that it affects the flow of protons and molecules such as ATP causing an imbalance in the membrane potential that causes the formation of pores through which vital components for the cell take place. Among the most susceptible microorganisms are Gram-negative bacteria, standing out *E. coli*, *S. enterica* (Miladi et al., 2016, Pateiro et al., 2021).

In another study, *Streptococcus pyogenes* were confronted at carvacrol concentrations between 0.53 mM and 1.05 mM with an instantaneous bactericidal effect on the pathogen. Carvacrol caused a destabilization of the membrane that caused the loss of cytoplasmic content, such as nucleic acids and enzymes such as lactate dehydrogenase. The impact on the permeability of the membrane was mainly due to the formation of pores, which causes the exit of ions such as K^+ and Na^+, which have a great impact on the electrical balance of the membrane and the functioning of enzymes essential for the functioning of the cell. This study was verified the synergistic effect of carvacrol with antibiotics, such as penicillin or clindamycin, to treat *S. pyogenes* infections. These results demonstrate the possibility of combining naturally occurring compounds such as carvacrol and traditional antibiotics as a therapy to treat diseases caused by pathogenic bacteria (Wijesundara et al., 2021).

Bouhhid et al. (2009), carried out a study with *P. aureuginosa* and *S. aureus* in which it was possible to detect affectations in the permeability to the K^+ ion in the cell membrane due to the interaction with carvacrol. After the treatment of cell suspensions with a concentration of 150 mg/L, the K^+ immediately began to leak out of the cell, reaching a volume of 0.09 ppm after 10 min of incubation with the antimicrobial. The same test was carried out using 225 mg/L, and the leakage of the ion of interest began immediately, gradually increasing with the incubation time until reaching 0.24 ppm after 2 h of contact. In *S. aureus*, K$^+$ efflux reached 0.64 and 0.77 ppm after 10 min of treatment with 150 and 225 mg/L, respectively. Along with the exit of K$^+$, the cells showed a considerable loss of membrane potential; probably because K$^+$ is an important factor in maintaining the balance between electrical charges through various mechanisms such as the Na$^+$/K$^+$ pump.

A study by Stratakos et al. (2018), with eight strains *E. coli* strains, including O157: H7, evaluated different concentrations of carvacrol and its effect on membrane permeability, intracellular ATP concentrations, and protein and nuclear material efflux. The presence of carvacrol caused severe damage to the electrical conductivity of the membrane, causing the exit of Na$^+$, K$^+$, and H$^+$ ions. The damage to the membrane caused by carvacrol is very severe, even allowing the exit of proteins and nuclear material to the cell outside. The decrease in the ATP intracellular concentration and its release to the external environment occurs through pores in the membrane. The rapid ATP depletion is linked to the uninterrupted hydrolysis by the membrane ATPase, responsible for keeping the proton flux stable. On the other hand, de Souza et al. (2013), found that lipophilic compounds can exert an antimicrobial effect on Gram-negative and Gram-positive bacteria causing abnormality on the cell or intracellular surface causing a flow of intracellular ATP towards the outside of the cells. This exit of ATP can be through pores formed in the membrane by the binding of carvacrol to it and cause irreparable damage to cell homeostasis. The depletion of ATP after the addition of lipophilic compounds results from their stimulation on the activity of ATPase of translation of protons and/or ions, which can originate from the elimination of the ionic driving force that limits the hydrolysis of ATP by ATPase.

The main consequence of the indiscriminate use of antibiotics is the increased resistance of bacteria to these compounds. One of the mechanisms that bacteria use to decrease the effect of antibiotics is the NorA antibiotic efflux pump. Efflux pumps, such as NorA, excrete antibiotics and other toxic substances to cells outside, reducing the effectiveness of the methods used to eliminate pathogens. For example, the case of *S. aureus*, a study developed by dos Santos Barbosa et al. (2021) demonstrated the effectiveness of carvacrol treatment for the inhibition of the NorA efflux pump in two strains of the pathogen. The combination of carvacrol with norfloxacin applied to the wild and multi-resistant strains achieved a decrease in MIC of 50% for norfloxacin. The efflux pump helps release substances toxic to the cell such as ethidium bromide. However, adding carvacrol achieved a 25% decrease in the MIC of this compound for the strains analyzed. Molecular docking studies confirmed that carvacrol acts as a competitive inhibitor, interacting with the amino acids that make up the active site of NorA and preventing the binding of norfloxacin and its excretion by pathogens.

Other reports of the effect of carvacrol show that adding 1 mM of this compound to cultures of *E. coli* O157: H7, causes the total inhibition of the development of the flagellum, which is very useful in the prevention of foodborne diseases. Flagella help bacteria in the adhesion and colonization process; therefore, the suppression of this bacterial structure would significantly impact the ability of this pathogen to colonize the intestine (Coccimiglio et al., 2016). For their part, Miladi et al. (2016), compared the effect of two concentrations of carvacrol on Gram-negative and Gram-positive pathogenic bacteria. The results show that carvacrol concentrations of 250 and 500 mg/L significantly affect antibiotic resistance due to its inhibitory effect on the mechanism flow of antibiotics across the cell membrane. The negative influence of carvacrol on the antibiotic efflux pump slowed the elimination of the antimicrobials, which increased the time of action of the antibiotic on the pathogen.

Most of the tests have been carried out *in vitro*, although there are studies that have evaluated the effect of carvacrol directly on the food matrix. For example in ready-to-eat sausages added with carvacrol and inoculated with *L. monocytogenes* strains, the effect of carvacrol on changes in permeability, depolarization, membrane structure, and respiratory activity was evidenced. The studies focused on elucidating the possible mechanisms of action by measuring the effects of carvacrol on the pathogen cells. The results obtained indicate that carvacrol inactivates bacteria through multiple targets that cause alteration of cell membranes, cause cell lysis, and inhibits respiratory activity. The addition of carvacrol had similar effects to the heat treatment at 70°C for 30 min, being detected degenerative changes in the membrane, mainly associated with changes in depolarization and permeability. Another harmful effect was the aggregation of genetic material and a significant decrease in respiratory activity (Churklam et al., 2020).

A study by Ma et al. (2022) analyzed the effect of carvacrol on two of the major pathogens in food: *E. coli* O157: H7 and *S. typhimurium* in chicken breast, a complex matrix very prone to bacterial contamination. One of the effects of carvacrol was the inhibition of biofilm formation in both pathogens, and the induction of reactive oxygen species, with a greater effect on *S. typhimurium*. Another effect detected is the deformation of the cell membrane that leads to its rupture and the leakage of nuclear material, cytoplasmic, and membrane proteins such as β-Galactosidase (Ma et al., 2022). These effects on the membrane and ROS production cause a significant decrease in the bacterial population compared with the untreated matrix after four days of incubation at 4°C.

Another important pathogen in meat and meat products that is sensitive to the action of carvacrol is Enterobacter cloacae (Zhuang et al., 2019; Sadek et al., 2021). In the study by Liu et al. (2021), the addition of concentrations of 64-256 µg/mL caused the inhibition of the biofilm formation of this pathogen *in vitro* studies the inhibition of biofilm formation could be correlated with a decrease in the exopolysaccharide production due to the presence of carvacrol. From transcriptional studies, it was found that carvacrol inhibits the expression of curli fimbriae genes between 2 and 9 times compared with type I fimbriae. Curli fimbriae are responsible for the adhesion of bacteria to colonized surfaces, which significantly influence the virulence of the pathogens. Another transcriptional effect was the inhibition of colonic acid biosynthesis and *fsrA*, *ftsQ* and *ftsZ* genes. These genes are related to cell division, while colonic acid is essential in the processes of aggregation, adherence, and maturation of the Enterobacteria biofilm (Sheng et al., 2020).

Essential oils are lipophilic compounds with a significant affinity for the cell membrane, capable of integrating into it, causing physicochemical changes in its properties. The interaction of the membrane with carvacrol affects the stability of the lipid bilayer, affects its integrity by forming pores that cause a flow of protons outside the cell. The pathogen tolerance to high antimicrobial concentrations is determined by their cellular structures capable of acting as barriers: the LPS layer (Gram-negative), the thickness of the peptidoglycan layer (Gram-positive), and the capsule. As has been reported in different studies, the mode of action of carvacrol as an antimicrobial agent is related to the involvement of several aspects of cellular homeostasis at the same time. *In vitro* studies have widely documented the use of this compound as an inhibitor of the growth of pathogenic microorganisms, serving as a basis for its successful application in fresh, cured, or ready-to-serve meat products.

Use of Carvacrol against Foodborne Pathogens

Bacteria that cause foodborne illness are a broad group of microorganisms, including various microbial species. Several methods are used to eliminate or contain foodborne illnesses, standing out the use of essential oils from plants or compounds derived from these, such as carvacrol. Being substances of natural origin, they have a better acceptance by the consumer and broader health benefits than the chemical compounds traditionally used in the food industry. The use of carvacrol as an antimicrobial agent that limits the development of pathogens in food, mainly in meat, has given excellent results, as described in the examples below.

The incorporation of 300 mg/L of carvacrol in a teriyaki sauce to marinate beef for 7 days at 4°C demonstrated the ability of carvacrol to inhibit and inactivate all contaminating native microbiota (aerobic mesophiles and total coliforms), in addition to inactivating the initial inoculum of 3.1 log CFU/g of *E. coli* O157: H7, *L. monocytogenes* and *S. tiphymurium* (Moon et al., 2017). In comparison, meat marinated in the sauce that did not contain carvacrol did not limit the growth of pathogens or other contaminating microorganisms. The practical application of carvacrol in this type of processed meat proved to be effective in removing contaminants only using natural compounds. It is a convenient methodology, where applying small amounts of carvacrol is achieved a significant bacterial inactivation. In addition, applying carvacrol directly to the product allows a continuous antibacterial effect during the storage, where the bactericidal effect is preserved during the distribution of the product (Moon et al., 2017).

In the study developed by Wang et al. (2021), the analysis of the *P. fluorescens* transcriptome showed that applying concentrations between 0.8 and 1.6 mM could significantly inhibit the production of exopolysaccharides that intervene in the formation of biofilms. Furthermore, the *luxI/luxR* genes are related to the production of acyl-homoserine lactones, molecules involved in quorum sensing, and biofilm formation. Interaction with carvacrol altered the transcription of genes associated with bacterial chemotaxis, flagellar assembly, and the citrate cycle. Other inhibited signaling pathways were exopolysaccharide synthesis, quorum sensing signaling pathways, and amino acid degradation, which may contribute to biofilm development and pathogen spoilage potential. In the same way, bacterial motility was affected, significantly decreasing swimming and swarming of cells, although there were no direct effects on cell viability.

Other studies showed that carvacrol has been used in food against pathogenic microorganisms. Juneja et al., (2012) combine the effect of this antimicrobial with a heat treatment to enhance the thermal death effect over *Salmonella* spp. in raw ground chicken. This study mixed up to 10 g/kg of carvacrol with the meat matrix, varying the cooking temperatures from 55 to 71°C. Carvacrol application significantly reduced cooking times; however, the study only presents results at the laboratory level. Analysis of the impact of adding carvacrol on an industrial scale would be advisable to determine the real benefits of using antimicrobials on a large scale. The major deficiency in this study is the lack of a sensorial analysis to determine if the applied carvacrol amount is effective against the pathogen and preserve the meat sensorial properties.

Another positive example of the use of carvacrol for the control of pathogenic microorganisms was evidenced by adding the antimicrobial to pieces of marinated chicken breast (Karam et al., 2019). *Brochothrix thermosphacta*, *Pseudomonas* spp., and *E. coli* are microorganisms that decreased their growth. They increased the product half-life by more than six days at 4°C using concentrations between 0.4 and 0.8% v/m of carvacrol. The combination of carvacrol in the marinade mix with vacuum packaging means a 3 log CFU/g decrease in the microorganisms that constitute the dominant microbiota in this food at the end of 21 days of treatment. Another outstanding effect reported in the study from the combination of carvacrol and vacuum is the extension of the sensory quality of the product from 15 to 21 days, compared to the untreated, which was only nine days.

Chicken ground meat is an ingredient widely used in the industry as a raw material to elaborate several products, and the use of carvacrol in this meat matrix has demonstrated its efficacy against pathogens such as *Salmonella* and *L. monocytogenes* (Chuang et al., 2020). In this work, the combination of a high-pressure treatment (350 MPa for 10 min), together with the use of carvacrol caused a significant decrease in both pathogens. Combining physical methods and natural antimicrobials is an excellent strategy to eliminate contaminating pathogens in meat matrices without affecting the product quality. In the case of carvacrol and other volatile compounds, their depletion occurs rapidly, influenced by their volatility, low solubility in water, or interaction with fats, factors that decrease its efficiency. Water/oil emulsions are a way to avoid the loss of carvacrol effectiveness in meat matrices; for example, its application in goat meat extends the carvacrol effect for nine more days regarding with a direct application (Syed et al., 2020).

The application of sulfites in meat products is a measure to prevent lipid oxidation and control pathogenic microorganisms such as *Pseudomonas*, *C. botulinum*, *L. monocytogenes*, *B. thermospacta*, Enterobacteria, and lactic acid bacteria. The application of carvacrol in lamb burgers demonstrated high efficiency in reducing the microbial contaminant load, with values similar to the traditional application of sulfites, using concentrations between 300 and 1000 ppm. The carvacrol antimicrobial effect remained constant throughout the trial period, with the most affected genera *Enterobacteriaceae* and *Pseudomonas* (Syed et al., 2020).

The application of carvacrol and other natural antimicrobials can be performed with different methods to guarantee a longer effectiveness time in the meat matrices and prevent factors such as composition, pH, water activity, or storage temperature from negatively impacting their effect on pathogens. Active packaging is a safe way to apply carvacrol in higher concentrations to avoid contamination with pathogenic microorganisms during the storage period since contamination begins on the surface of the food (Pateiro et al., 2019). This strategy does not imply a modification of the sensory attributes, one of the main difficulties when applying carvacrol or other essential oils in meat products. Aytac et al. (2017) obtained a significant reduction of *C. perfringens* in sausages, similar to that obtained with traditional chemical additives, from the application of gelatin and chitosan nanocapsules with carvacrol and thymol. Nanocapsules prolong the antimicrobial effect of essential oils over time without affecting the sensory attributes of the product.

Just as the antimicrobials of natural origin are added to food to limit the pathogenic microorganisms' development without affecting the functional and sensory properties within the components of the product, some food components can serve as a barrier or help reduce the effect of these antimicrobials. For example, food has in its composition all the nutrients necessary for the development and multiplication of microorganisms, therefore, the interaction with any of these components such as proteins or lipids can serve microorganisms to limit the action of additives and protect them of their mechanisms of action. Other physical-chemical characteristics of the product such as pH, water activity, or temperature can accelerate the volatility of the compounds and minimize their antimicrobial effect (Kachur and Suntres 2020; Pateiro et al., 2021).

Factors Affecting the Antimicrobial Action in Meat Matrices

There is evidence of the interaction of essential oils and their components with the different compounds of food matrices. Fat, protein, and water influence the solubility of carvacrol and other terpenes, which can change with the pH of the medium. High concentrations of fat in meat can interfere with the antimicrobial activity of carvacrol and other terpenes, as reported by Vasan et al. (2014). The amount of fat can inhibit the antimicrobial effect of carvacrol on *E. coli* O157: H7 and *Salmonella* in ground beef and sausage dough. The interaction between the different components can even affect the thermal resistance of the pathogens, increasing the thermo-tolerance, increasing the D value, and causing the thermal treatments to require higher temperatures to achieve the elimination of the pathogens.

The application of carvacrol in biofilms has given good results, managing to maintain the antimicrobial properties despite not being in direct contact with the meat matrix. In the research developed by Wang et al. (2020) lactic acid biofilms that included carvacrol were tested in ground beef with different fat concentrations. In the matrix with the fat concentration of 12%, the adsorption of carvacrol was 1.3 times higher than that containing only 5%. Another interesting effect found is that although the permeability of carvacrol increases as the percentage of fat increments, the highest antimicrobial effect was observed in lean meat. The impact of relative humidity was not significant, although this parameter varied from 94 to 43% over 12 days. The temperature had a significant effect on the carvacrol migration towards minced meat since temperatures augmentation (from 5 to 30°C), increase the amount of released carvacrol from the films.

de Souza et al. (2020) evaluated the effect of carvacrol in conjunction with montmorillonite (complex polymer) in starch films to inhibit the growth of *E. coli* in fresh products. The result showed that the interaction between montmorillonite and starch created a more temperature-resistant structure that allowed carvacrol to remain within it for a longer time. In bacterial cells, a destabilization and partial destruction of the cell membrane occurred due to the interaction with carvacrol. The film structure that presented the least crystallization allowed the best migration of the carvacrol out of the film and achieved the highest bactericidal effect.

There are different matrices derived from whey proteins used to obtain films and coatings for meat products, and when mixed with carvacrol achieve a significant reduction of pathogens such as *E. coli* and *Pseudomonas* spp (Šimat et al., 2021). In marinated products such as pork and salmon fillets with 1% of carvacrol incorporated, this antimicrobial had a bactericidal effect on the entire contaminating microbiota (*E. coli*, coliforms, lactic acid bacteria, fungi, and yeasts). Ingredients such as NaCl (10% w/w), sodium lactate (2%), and Tween 80 (emulsifier) were used in the sauce to the marinade, with a final pH of 4.5. The bactericidal effect found in the marinade sauce is attributed to carvacrol since the sauce alone did not demonstrate any antimicrobial effect. However, the high content of salt and organic acids had no impact on the antimicrobial activity of the carvacrol added to the mixture (Van Haute et al., 2016).

The interaction of carvacrol with some components of meat products such as salt (NaCl) has a synergistic effect on bacterial growth inhibition. This effect can be associated with the combination of the mechanism of action of both compounds; carvacrol causes damage in the cell membrane, while NaCl creates an osmotic imbalance that causes cells death. In the work developed by (Kim et al., 2020), the salt and carvacrol combination resulted in a significantly higher reduction of *E. coli* O157: H7, *L. monocytogenes*, and *S. aureus* compared to the individual compounds. The combination of carvacrol (2%) and NaCl (> 3%) eliminated all the initial microbial load (7.2–7.4 CFU/mL) after one minute of interaction with pathogens. This study shows that the decreased NaCl content used as an antimicrobial could be replaced by compounds of natural origin, such as carvacrol, with superior results in reducing the microbial load of foods.

Despite the wide use of carvacrol in meat products, it is not very common to evaluate the effect of food components such as fat or pH on antimicrobial properties on pathogens such as *S. typhimurium* PT4 and *E. coli* O157: H7. Carvacrol has a higher solubility in organic compounds such as fat, so the amount of fat in meat can affect its distribution in the matrix, as well as its antimicrobial activity. In the case of *S. typhimurium*, the inhibitory effect was affected by decreasing fat and pH. In the case of *E. coli*, the antimicrobial effect of carvacrol was lower in the treatments that had the highest fat concentration and the highest pH. This study reports a significant interaction effect between fat and pH that interferes with the antimicrobial effect of carvacrol. The impact of fat is related to the lower availability of carvacrol in the aqueous phase where most of the microorganisms are present, and the pH tested (pH 6) limits the affinity of carvacrol for fat, making it maintain its inhibitory effect on the pathogens. Finally, fat promotes changes in the extracellular environment, increasing the interaction of carvacrol with cell membranes, enhancing its growth inhibitory effect (Carvalho et al., 2018).

The interaction between meat matrices components, carvacrol, and other antimicrobials can affect or enhance their activity against pathogens. The correct formulation of the products becomes a challenge to balance quality and safety, providing consumers with excellent products free of microbial contaminants. Similarly, high concentrations of carvacrol and other essential oils can negatively impact quality aspects such as color, odor, and taste, dulling their antimicrobial effect. In addition, performing sensory studies is an excellent complement to antimicrobial activity that should be considered.

Thermal Processing of Meat and Meat Products: D-Value and z-Value

Meat and meat products are high nutrient foods and an excellent means of pathogens propagation, such as *L. monocytogenes*, *E. coli*, and *Salmonella*, causing numerous food outbreaks when the products are not correctly cooked (Lahou et al., 2015). Heat treatment is the main method applied to the foods used to eliminate foodborne pathogens. Thermal processing of food involves cooking temperatures ranging from 50 to 150°C for defined times. This process seeks to guarantee the microbiological safety of the product, extend its shelf life, and improve its physicochemical and sensory characteristics. Fruits and vegetables are more sensitive to the action of heat, while meat and meat products generally need extended exposition time to achieve appropriate physicochemical and sensory characteristics. USDA recommends cooking meat and meat products at an internal temperature of 71.1°C measured using a thermometer in the center of the food. However, consumers generally use the visual color of the food to judge the doneness, even though this does not guarantee food safety (Lin 2018). It has been reported that, at an internal cooking temperature of 57°C, the visual color of meat and some meat products are similar to a fully cooked, but insufficient to eliminate foodborne pathogens (Lyon et al., 2000). Hague et al. (1994), cooked beef patties at an internal temperature of 55 and 77°C. Results show that some patties cooked to lower temperatures turned brown prematurely, and concluded that visual evaluation of internal patty color was not an accurate indicator of patty doneness.

Meat and meat products are processed using different methods, such as sous vide used for prolonged low-temperature treatments. The sous-vide method consists of introducing raw or precooked food into a heat-resistant container, extracting the air from inside, sealing it hermetically, and finally subjecting it to a constant temperature for the specified time (Rinaldi et al., 2014). The basic material and equipment required for sous vide cooking included a thermometer, vacuum sealer, water oven or thermal circulator, and vacuum pouches. The low temperature of about 60°C or lower applied in this method (Abel et al., 2020) is in the borderline of temperature where some pathogens could grow; thus, inappropriate sous vide cook process could compromise the microbiological food safety. Some studies reported in the literature evidenced the microbiological risk of foods inappropriately processed under the sous-vide method. Jørgensen et al. (2017) evaluated the microbiology quality of lightly cooked food, including sous-vide, at the point of consumption in England. The authors analyzed the samples for the presence of *Campylobacter* spp., *Salmonella* spp., *Bacillus* spp., *C. perfringens*, *Listeria spp.*, *S. aureus*, and *E. coli*. Results were interpreted as unsatisfactory, borderline or satisfactory, and the percentage of samples sous vide processed in each range were 29, 24, and 47%, respectively. The proportion of samples of an unsatisfactory microbiological quality found in food prepared using sous-vide was the highest between the different cooking methods evaluated.

The sous vide process is very safe if applied correctly, but an inappropriate thermal process can compromise the safety of the product and can lead to a food outbreak. McIntyre et al. (2017) investigated some illnesses by *Salmonella* due to the consumption of sous vide cooked foods, one of them linked to restaurants serving sous vide duck breast expended in British Columbia, Canada. Researchers suggested that the most probable cause of the illness was improper sous vide cooking practices.

Cooking meat and meat products taking into account only the visual meat color or implementing inadequate thermal processes using the sous-vide method are just two examples of process deviations representing a risk to the consumer. Implementing an adequate heat treatment is essential to ensure food safety. The D and z values are important parameters for implementing adequate heat treatments during the preparation of meat and meat products. The D value is the time in min required to inactivate 90% of the viable cells of a particular microorganism at a specific temperature and is determined from the exponential linear portion of the inactivation curve. The number of survivors is quantified with respect to time for each heating temperature and the D-value is calculated as the negative inverse slope of the plot. On the other hand, the z-value is defined as the increase in temperature in °C needed to change the D value by one logarithmic unit and is calculated by linear regression of the log10 D values vs. their corresponding heating temperatures. The z-value is calculated by taking the absolute value of the inverse of the slope and is used to measure the resistance of the pathogen to changes in temperature.

The D value informs about the resistance of a microorganism at a specific temperature and the minimum process time required to guarantee food safety. The D value is specific for each microorganism and dependent on the meat matrix and food composition as shown by some researchers. Osaili et al. (2007), carried out a study with pork patties determining the D value at temperatures between 55-70°C for *Salmonella* spp., *L. monocytogenes*, and *E. coli* O157: H7. After estimating the D values, the results showed that *E. coli* O157: H7 was the least resistant, obtaining a D value of 0.08 min at 70°C, while *L. monocytogenes* was the most resistant with a D value of 0.43 min at

70°C. Abel et al. (2020) investigated the effects of sous vide cooking at 50, 55, and 60°C on the thermal resistance of *L. monocytogenes* inoculated in roe deer and wild boar meat results showed that the D-values were largely dependent on the meat matrix. D-values for *L. monocytogenes* in roe deer meat at 50, 55, and 60°C were 49.2, 14.9, and 3.7 min, respectively, whereas the wild boar reached D-values of 100.2, 23.8, and 4.2 min.

On the other hand, Juneja and Eblen (2000) reported a protective effect of fat on the heat resistance of *S. typhimurium* in ground beef. The pathogen heat resistance was evaluated in ground beef containing 7, 12, 18, and 24% fat. According to their results, contaminated ground beef containing 7% fat should be heated to an internal temperature of 65°C for 7.1 min to achieve a thermal process of 7-log reduction. However, when fat content increased to 24%, the time increased to 20.1 min to reach the same log reduction. Ahmed et al. (1995), reported similar results on the heat resistance of *E. coli* O157: H7 in ground beef, pork sauce, and chicken and turkey meat affected by the fat level. For all evaluated products, higher fat levels resulted in higher D-values.

The USDA (2001) published mandatory compliance guidelines to help food processors meet standards and establish critical limits for cooked meat. The norm provides internal temperature and time combinations to achieve the mandatory 6.5 log reduction of *Salmonella* and 5.0 log for *E. coli* O157: H7 in beef and ready-to-eat products. These guidelines apply to beef products and those derived from other animal species, such as poultry and pigs. The thermal processes defined in these guidelines are sufficient to ensure the inactivation of foodborne pathogens and provide consumer confidence that the food has been properly processed and is safe for consumption.

The Effects and Interactions of Carvacrol, Fat, and Temperature on the Thermal Inactivation of Pathogens in Meat Products

The incorporation of natural antimicrobials in food is an excellent strategy to increase the safety margins of food during cooking. For example, carvacrol is a natural compound with antimicrobial properties that can be used in meat formulations. The antimicrobial action of carvacrol is based on its interaction with the cell membrane, causing irreparable damage to the structure by increasing permeability and depolarization (Ait-Ouazzou et al., 2013). It has been observed that adding carvacrol to the growth medium of *E. coli* causes an increase in the flux of PO_4^- and K^+ ions, from the inside to the outside of the cell membrane (Can Baser 2008). In addition, carvacrol is used in synergy with heat to increase safety margins during the inactivation of *Salmonella* in chicken meat (Juneja et al., 2012), as well as *E. coli* O157: H7 and *C. perfringes* in ground beef (Juneja and Friedman 2008; Juneja et al., 2006).

Fat in meat products plays an important role in the texture and sensory properties, and it also affects the survival of pathogenic microorganisms by acting as a barrier against heat during heat treatments (Juneja et al., 2001). The meat matrix forms fat micelles that act as a barrier and can contain pathogenic bacteria such as *E. coli* O157: H7. Heat transmission becomes heterogeneous, and isolated thermal points favor bacterial resistance, reducing the effectiveness of heat treatments. Bacteria suspended in fat are harder to destroy than those in an aqueous medium due to a reduction in water activity. An increase in the fat content decreases the moisture content, altering the heat transfer (Ahmed et al., 1995). Therefore, as the fat content of foods increases, more attention should be paid during cooking to avoid the survival of pathogenic microorganisms.

The effect of temperature in the inactivation of foodborne pathogens is mainly due to the heat influence on cell membranes and proteins. The membrane configuration is affected because the phospholipids that form the bilayer change their structure, making it more fluid and more permeable to the components of the medium. Proteins embedded in the cell membrane, and cytosol, are also denatured, causing severe damage to membrane integrity and cell metabolism. Studying the combined effects of carvacrol, fat content, and heat on the resistance of pathogens can become a complicated task if the proper methodological strategy is not used. Prediction models are practical tools that facilitate the study of multiple factors that affect the thermal resistance of pathogens. The response surface methodology is a mathematical and statistical tool used to evaluate several factors simultaneously and estimate their linear, quadratic, and interaction effects (Montgomery 2017). Our research group studied the thermal resistance of *E. coli* O157: H7 in ground beef as a function of temperature and concentrations of carvacrol and fat using the response surface methodology. The resulting regression equation predicting thermal death (D value) of *E. coli* O157: H7 was:

$$Ln(D\text{-value}) = 0.097 - 1.25T + 0.18F - 0.13C - 0.1FC - 0.16T2 + 0.12F2$$

Where T = temperature, F = fat, C = carvacrol.

According to the model regression coefficients, the temperature (T = –1.25) was the most influential factor in the thermal resistance of *E. coli* O157: H7, followed by fat (F = 0.18) and carvacrol (C = –0.13). The interaction effects between fat and carvacrol (FC) and the quadratic effects of temperature (T2) and fat (F2) were also significant. The negative regression coefficients for temperature and carvacrol indicate the D value decreases and these increase. On the contrary, the addition of fat to ground meat shows an opposite behavior (positive regression coefficient), regarding the D value estimated by the model. These behaviors can be observed graphically in Figure 1a and Figure 1b where the combined effects of temperature-fat and temperature-carvacrol are shown. The quadratic effect of temperature is evidenced in these same figures by the curvature observed around 61°C, a temperature above which *E. coli* O157: H7 does not offer resistance to heat.

Fat has a protective effect against heat inactivation of *E. coli* O157: H7 in ground meat (F = 0.18), an increase in it causes the same trend in the D value. The fat effect on the D value is only at temperatures below 60°C from this value there is no influence on the resistance of the microorganism to heat, regardless of the amount of fat added (Figure 1a). From 60°C, the temperature has a greater effect on the pathogens' mortality that cannot be affected by the fat in the meat matrix. The melting temperature of beef fat is between 40 and 48°C, depending on the area from which it is obtained. At 60°C, the fat is in a liquid-state and its protective effect is reduced with respect to temperature because heat transmission is much faster than in the solid-state and the fat micelles that protected bacteria from the heat of cooking have been lost (Grompone 1991).

Regarding the effect of carvacrol on the D value, this presents a significant effect ($p < 0.05$), and according to the regression coefficient of the model (C = –0.13), the negative sign indicates that when carvacrol is added to meat, the D value of *E. coli* O157: H7 decreases. The influence of carvacrol depends on the amount of fat in the product as indicated by the significant effect ($p < 0.05$) of the fat-carvacrol interaction (FC = 0.1), this behavior is clearly seen in Figure 1c. At low-fat concentrations, around 5%, the effect of carvacrol is not as marked as at high concentrations. This tendency is due to the chemical properties of carvacrol that make it more soluble in nonpolar compounds since it has a partition coefficient LogP = 3.49 (Nazzaro et al., 2013). A high value of LogP (> 2.5) indicates that a substance is mainly hydrophobic. Therefore, carvacrol is more soluble in organic substances such as fat than in water (Baluja et al., 2017).

The regression model describes the behavior of the thermal resistance of *E. coli* O157: H7 as a function of the temperature (55-65°C), and fat concentration (5-20%) and carvacrol (0-2%) in ground beef. The food processor can select any combination of temperature, fat, and carvacrol, within established limits, to estimate the log reduction of *E. coli* O157: H7 in ground beef.

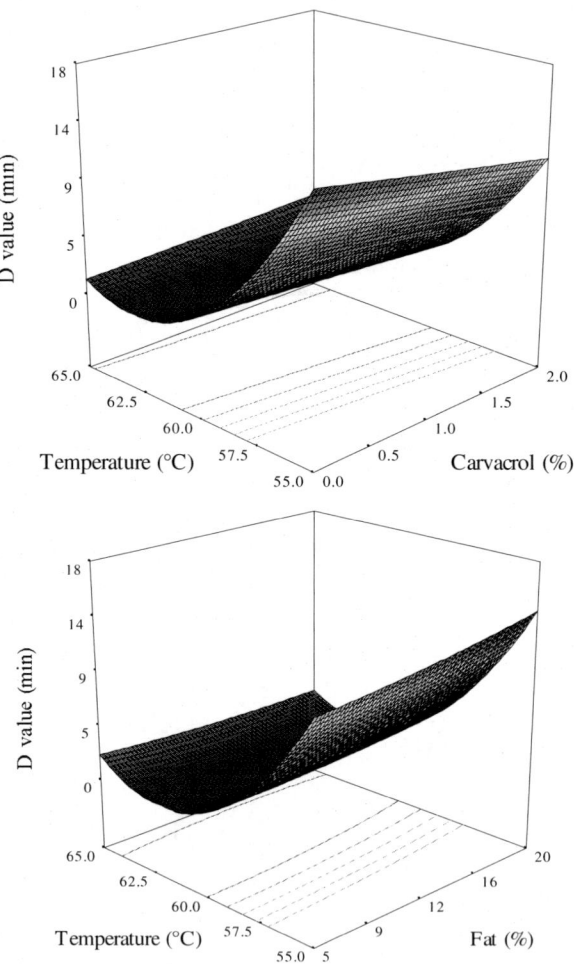

Figure 1. (Continued)

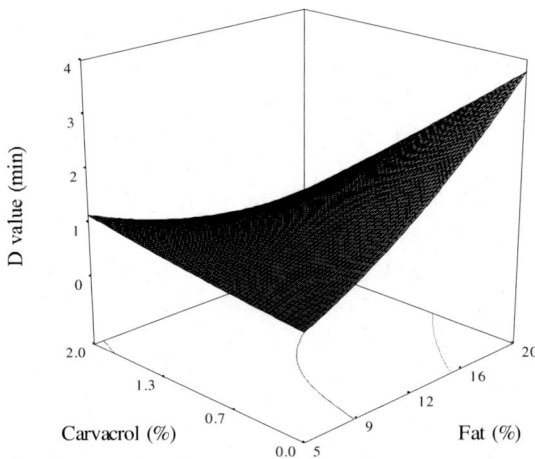

Figure 1. Effects and interactions of temperature, fat and carvacrol on the D value of *E. coli* O157: H7 in ground beef. A) temperature vs fat (carvacrol = 1.0%), B) temperature vs carvacrol (fat = 12.5%), C) fat vs carvacrol (temperature = 60°C).

Sensory Impact of Carvacrol on Meat Products

One of the main obstacles when applying natural antimicrobials is the negative impact on the sensory qualities of meat products. However, this impact can be mitigated using methodologies to attenuate the flavors and odors of these compounds in their pure state. Natural antimicrobials also are added to other foods such as fruit juices and nectars, where its flavors, although not usual, are well masked. Microcapsules containing essential oils are used in nectars and chopped fruits, which gradually release these compounds. In addition, these antimicrobials can be integrated into films and coatings, to prevent product contamination while maintaining sensory properties (Del Nobile et al., 2012). Neutral tasting substances like maltodextrin, alginate, or chitosan are used for encapsulation, making it a cheap and easy protocol. Another methodology reported in meat products is using biosurfactants to create micelles with microcapsules of essential oils. Gaysinsky et al. (2005) applied this method with carvacrol and eugenol to inhibit the growth of *E. coli* O157: H7 and *L. monocytogenes* in a sausage-type sausage, demonstrating its efficacy and feasibility.

The addition of essential oils as an antimicrobial in meat products is common, and some studies evaluate prediction models. (Juneja and Friedman

2008), used concentrations of 1% to analyze the model; however, the authors do not perform a sensorial analysis to corroborate the acceptance of the concentrations considered optimal for microbial inactivation in the prediction model. In another study carried out in 2012 (Juneja et al., 2012), applied 1% carvacrol in chicken meat to evaluate its effect on thermal resistance in *Salmonella*, without doing sensory studies to analyze the influence of this antimicrobial on the sensorial properties of the product. This type of study has become increasingly obsolete, due to the impact that essential oils have on the sensory aspect of food. Incorporating sensorial studies into trials provides a more realistic, multidisciplinary approach that helps meet consumer needs.

In a study developed by (Bellés et al., 2019), the sensorial panel that evaluated grilled lamb was able to detect the presence of carvacrol in the treatments with 300 and 1000 ppm, clearly distinguishing the differences in concentration between both. The herbaceous taste and smell caused the scores of these treatments to be lower than the control from 3 to 8 days after packaging the meat. The effect on the sensorial quality of lamb meat was significantly negative due to the addition of carvacrol; however, the antimicrobial effects were positive and promising. In this study, the relationship between the antimicrobial effect and the sensory impact of the concentrations used was negative. A similar study with lamb meat with 0.6 and 0.9% oregano essential oil, showed a high organoleptic acceptance, with better results for the lower concentration. Similarly, in bologna added with a low concentration (0.02%) of carvacrol, the panelists, despite detecting the aroma of oregano, the sensory quality was similar to the control. In other matrices such as rainbow trout, the addition of oregano essential oil (0.2%) resulted in better aromas and flavors in this product (Rodriguez-Garcia et al., 2016).

The combination of carvacrol with other antimicrobial compounds or methodologies are strategies to mitigate the strong flavor and odor of carvacrol in foods. The combination of methods generates a synergistic effect that enhances the bactericidal activity, maintaining the sensorial acceptance of the food. The combination of carvacrol with technologies such as vacuum packaging has a synergistic effect in reducing the antimicrobial resistance of pathogens and allows the use of lower concentrations. The application of treatment with 450 MPa reduces 90-99% of the *L. monocytogenes* cells in the meat, the addition of carvacrol enhances this pressure effect, achieving the complete elimination of the contaminating bacterial load from the matrix. The combination of both treatments has also been shown to be effective for eliminating *B. cereus* spores in heat treatments with temperatures ≤ 65°C (Li and Gänzle 2016).

The study by Karam et al. (2019) made an in-depth analysis of the sensory impact of adding carvacrol to marinated chicken meat. The aroma is one of the main attributes affected when adding high carvacrol concentrations to meat products. The results indicated that the low concentrations of the antimicrobial managed to extend the life of the product with high sensory quality for 21 days, much higher than the controls. The carvacrol treatments that showed the highest sustainability over time were the marinated chicken samples with the lowest concentrations (0.4%). The combination of carvacrol with vacuum packaging, low storage temperatures, and anaerobic conditions extend the product shelf life between 6 and 12 days. The use of carvacrol with levels lower than acceptable sensory values can act as an additional barrier that increases the lethality rate on pathogens without affecting the product´s quality (de Oliveira et al., 2019).

The sensory impact may be different when it is used the essential oil or dried herbs because the carvacrol concentration in both is different. However, the study by Jaworska et al. (2021) showed no effect on the sensory quality of ground chicken meat until 10 days of storage at 4°C. The addition of dried oregano had better results in preserving the sensory quality of chicken meat than essential oil, which causes a marked decrease in sensory acceptance. The unpleasant flavors that were most reported in the treatments were the herbaceous smell and the bitter taste, mainly in the treatments with the dry herb. Therefore, the type and quantity of essential oils with antimicrobial properties that can be used for meat products preservation are limited due to the acceptability of the final product. The current challenge facing the meat industry with the use of essential oils is to reduce their concentration without compromising their antimicrobial efficacy. The strategy has focused on using active compounds such as carvacrol since they require lower amounts to have a significant antimicrobial effect without sensory affecting the products.

It is important to choose the correct combination of desirable properties in essential oils and vegetal antimicrobial compounds to be added to meat products. The parameters of innocuousness, quality, safety, and flavor of meat products must be observed, when new additives or technologies are used to prevent the product from losing any of its main attributes, either due to the action of pathogens or due to compounds we add. Although antimicrobials of natural origin have proven to be highly effective in reducing the number of pathogens that affect meat and its derivatives, the odors or flavors that these contribute to the final product can affect the sensory quality. The essential oils of aromatic plants used as antimicrobials generally have a high degree of purity which influences the flavor of the product to which they are added, causing customer dissatisfaction. There are new technologies such as microencapsulation, the use of smart packaging, or the Response Surface Methodology, which allow a comprehensive analysis of a greater number of factors and contribute to the correct application of these compounds in food.

Application Prospect of Carvacrol in the Meat Industry

Carvacrol is a compound with several applications in the meat industry due to its antimicrobial and antioxidant properties. Bosetti et al. (2020), evaluated the effect of microcapsules of carvacrol and cinnamaldehyde on the weight gain, development, and meat quality of broilers. The researchers tested concentrations ranging from 30 - 400 mg/kg of an essential oil blend containing 30% carvacrol. The evaluated birds presented differences in the color of the meat; but they showed better development than the control group supplemented with virginiamycin, with the same yield and quality of meat. This study shows that essential oils such as carvacrol can be an excellent alternative to conventional antibiotics used as growth promoters in poultry.

Another similar study conducted by Ma et al. (2022), evaluated the joint effect of carvacrol, cinnamaldehyde, monocaprine, citral, and thymol on the shelf life of chicken meat inoculated with *E. coli* O157: H7 and *S. typhimurium*. The authors report that the joint application of carvacrol with monocaprin shows the best antimicrobial effect and causes the highest inhibition of bacterial biofilms. Monocaprin is a monoglyceride derived from capric acid with antimicrobial properties against bacteria, yeasts, and viruses. On the other hand, it has been suggested that carvacrol forms pore in lipid membranes, increasing the permeability of the membrane and affecting the cellular ionic balance (Marchese et al., 2018).

Ma et al. (2022), report that while carvacrol acts at the level of the outer membrane, monocaprin diffuses into the cell interior and increases the production of reactive oxygen species. Carvacrol increased the impact of monocaprin, and according to the authors, this synergistic effect is the cause of the significative inhibition of the growth of pathogens when applying this treatment. The authors also report that carvacrol and monocaprin application does not affect the sensorial properties of the product. On the contrary, this combination improves product acceptance by extending the meat shelf life stored at 4°C. It is necessary to highlight that the study carried out by Ma et al. (2022), was realized on chicken breast, a matrix with a low-fat content that, therefore, does not offer protection to microorganisms. However, the results reported by the authors highlight the possibility of applying carvacrol and its combination with other bioactive compounds to prolong the shelf life of meat products (López-Pino et al., 2021).

A study conducted by Bellés et al. (2019), compared the benefits of the application of carvacrol and green tea essential oils with those presented when using sulfites in lamb meat for hamburgers. In this study, the authors reported the changes in the bacterial populations of meat, in response to the individual or combined effect of oils. Bellés et al. (2019), reported the oils' antioxidant action by evaluating the oxidation of lipids and polyphenols, and the organoleptic qualities analysis. The authors found that essential oils are a better alternative for preserving meat since, with the application of 300 ppm of carvacrol or green tea, lower lipid oxidation is obtained compared to the use of 400 ppm of sulfite. An outstanding result of this study is that only carvacrol inhibited the growth of the microorganisms. However, one thing to consider is that carvacrol provides an herbal smell and taste; affecting the general acceptance of the product by consumers. Therefore, although carvacrol alone is an attractive alternative for food preservation, the method of application and concentrations are important aspects that must be studied to maintain the sensory characteristics of the meat.

The microencapsulation of essential oils is another of the new methodologies that are being applied to preserve food. Jouki and Khazaei (2021) studied the impact of a coating made with quince seed (QSG) and carvacrol microcapsules (CM) on the organoleptic and physicochemical properties of chicken nuggets. In this study, the authors mixed the active ingredients (QSG and CM) with the powder to bread the nuggets, measuring the parameters of interest before and after frying. Jouki and Khazaei (2021), report that active coatings reduced water loss and oil absorption by approximately 30%. The authors report that the encapsulated carvacrol maintained its antioxidant activity and the coated samples showed the lowest oxidative stress markers. Additionally, the sensory analysis of the samples with the quince seed coating and carvacrol microcapsules did not show differences from the control samples. Overall, the results of this study show that carvacrol can be used in processed products, taking advantage of its antioxidant power without affecting the sensory quality of the product. Finally, the use of methods such as microencapsulation or active packaging is feasible alternatives for the use of carvacrol and other essential oils in food preservation.

Conclusion

As discussed throughout this chapter, carvacrol is a compound with widely demonstrated antimicrobial properties against major food pathogens such as *E. coli*, *S. typhimurium*, *L. monocytogenes*, and others. The mechanisms of action have proven to be very diverse, mainly affecting cell membranes and ionic balance, achieving a significant decrease in the microbial load in different food matrices. The interaction with the components of meat products such as fat or salt modulates the antimicrobial effect of carvacrol due to its hydrophobic nature. The versatility of carvacrol allows its use in different ways, whether added directly to the product due to its spicy flavor (oregano smell and taste in sauces and marinades) or as an integral part of coatings and films that work as a barrier to limit the access of microorganisms to food. In the technological aspect, the use of carvacrol in meat products affects cooking times and temperatures since it can modify the D value of pathogenic microorganisms, increasing their sensitivity to the effect of heat treatment.

In several meat products it has been proven that adding carvacrol in low concentrations (<1%), its impact on the thermal resistance of microbial contaminants makes it possible to apply lower cooking temperatures or for shorter times, with the consequent benefit for the quality nutritional value of food. The addition of essential oils or their active compounds has been shown to affect sensorial food quality, for which reason their application has been limited to some sectors and products. However, using new techniques and methodologies such as spray drying, microencapsulation, and slow-release films and coatings may help carvacrol be used more and more in a greater number of foods. The use of natural antimicrobials has become a premise in the food industry, where more and more consumers prefer foods with fewer chemical additives and better nutritional properties with the highest level of safety and quality. Carvacrol is a compound that meets the premises sought by the modern food industry, providing security due to its antimicrobial character and added value due to its antioxidant properties, which are widely desired for the meat industry, for which in the short terms its use in meat and meat products will undoubtedly increase.

References

Abel, Tobias, Annika Boulaaba, Karolina Lis, Amir Abdulmawjood, Madeleine Plötz. & André Becker. (2020). "Inactivation of *Listeria monocytogenes* in game meat applying sous vide cooking conditions." *Meat science*, no. *167*, 108164. doi: 10.1016/j.meatsci.2020.108164.

Agrimonti, Caterina, Jason C White, Stefano Tonetti. & Nelson Marmiroli. (2019). "Antimicrobial activity of cellulosic pads amended with emulsions of essential oils of oregano, thyme and cinnamon against microorganisms in minced beef meat." *International journal of food microbiology*, no. *305*, 108246. doi: 10.1016/j.ijfoodmicro.2019.108246.

Ahmed, Nahed M., Donald E Conner. & Dale L Huffman. (1995). "Heat-resistance of *Escherichia coli* O157: H7 in meat and poultry as affected by product composition." *Journal of Food Science*, no. *60* (3), 606-610. doi: 10.1111/j.1365-2621.1995.tb09838.x.

Ait-Ouazzou, A, Espina, L., Gelaw, T. K., de Lamo-Castellví, S., Pagán, R. & García-Gonzalo, D. (2013). "New insights in mechanisms of bacterial inactivation by carvacrol." *Journal of applied microbiology*, no. *114* (1), 173-185. doi: 10.1111/jam.12028.

Aminzare, Majid, Mohammad Hashemi, Hassan Hassanzadazar. & Jalal Hejazi. (2016). "The use of herbal extracts and essential oils as a potential antimicrobial in meat and meat products; a review." *Journal of Human, Environment and Health Promotion*, no. *1* (2), 63-74. doi: 10.29252/jhehp.1.2.63.

Aytac, Zeynep, Zehra Irem Yildiz, Fatma Kayaci-Senirmak, Turgay Tekinay. & Tamer Uyar. (2017). "Electrospinning of cyclodextrin/linalool-inclusion complex nanofibers: Fast-dissolving nanofibrous web with prolonged release and antibacterial activity." *Food Chemistry*, no. *231*, 192-201. doi: 10.1016/j.foodchem.2017.03.113.

Baluja, Shipra, Anchal Kulshrestha, & Jagdish Movalia. (2017). "1-Octanol-water partition coefficient of some cyanopyridine and chalcone compounds." *Revista Colombiana de Ciencias Químico-Farmacéuticas*, no. *46* (3), 342-356. doi: 10.15446/rcciquifa.v46n3.69466. [*Colombian Journal of Chemical-Pharmaceutical Sciences*]

Bellés, Marc, Veronica Alonso, Pedro Roncalés. & Jose A Beltrán. (2019). "Sulfite-free lamb burger meat: antimicrobial and antioxidant properties of green tea and carvacrol." *Journal of the Science of Food and Agriculture*, no. *99* (1), 464-472. doi: 10.1002/jsfa.9208.

Bosetti, Gilnei E., Letieri Griebler, Edemar Aniecevski, Caroline S Facchi, Cintiamara Baggio, Gabriel Rossatto, Felipe Leite, Fernanda DA Valentini, Alicia D Santo. & Heloisa Pagnussatt. (2020). "Microencapsulated carvacrol and cinnamaldehyde replace growth-promoting antibiotics: Effect on performance and meat quality in broiler chickens." *Anais da Academia Brasileira de Ciências*, no. *92*, doi: 10.1590/0001-3765202020200343. [*Proceedings of the Brazilian Academy of Sciences*]

Bouhdid, S., Abrini, J., Zhiri, A., Espuny, M. J. & Manresa, A. (2009). "Investigation of functional and morphological changes in Pseudomonas aeruginosa and *Staphylococcus aureus* cells induced by *Origanum compactum* essential oil." *Journal of Applied Microbiology*, no. *106* (5), 1558-1568. doi: 10.1111/j.1365-2672.2008.04124.x.

Can Baser, K. H. (2008). "Biological and pharmacological activities of carvacrol and carvacrol bearing essential oils." *Current pharmaceutical design*, no. *14* (29), 3106-3119. doi: 10.2174/138161208786404227.

Carvalho, Rhayane I., Andrea S de Jesus Medeiros, Maísa Chaves, Evandro L de Souza. & Marciane Magnani. (2018). "Lipids, pH, and their interaction affect the inhibitory effects of carvacrol against *Salmonella typhimurium* PT4 and *Escherichia coli* O157: H7." *Frontiers in microbiology*, no. *8*, 2701. doi: 10.3389/fmicb.2017.02701.

Coccimiglio, John, Misagh Alipour, Zi-Hua Jiang, Christine Gottardo. & Zacharias Suntres. (2016). "Antioxidant, antibacterial, and cytotoxic activities of the ethanolic *Origanum vulgare* extract and its major constituents." *Oxidative medicine and cellular longevity*, no. *2016*, doi: 10.1155/2016/1404505.

Chuang, Shihyu, Shiowshuh Sheen, Christopher H Sommers, Siyuan Zhou. & Lee-Yan Sheen. (2020). "Survival evaluation of *Salmonella* and *Listeria monocytogenes* on selective and nonselective media in ground chicken meat subjected to high hydrostatic pressure and carvacrol." *Journal of food protection*, no. *83* (1), 37-44. doi: 10.4315/0362-028X.JFP-19-075.

Churklam, Wasinee, Soraya Chaturongakul, Bhunika Ngamwongsatit. & Ratchaneewan Aunpad. (2020). "The mechanisms of action of carvacrol and its synergism with nisin against *Listeria monocytogenes* on sliced bologna sausage." *Food Control*, no. *108*, 106864. doi: 10.1016/j.foodcont. 2019.106864.

de Medeiros Barbosa, Isabella, José Alberto da Costa Medeiros, Kataryne Árabe Rimá de Oliveira, Nelson Justino Gomes-Neto, Josean Fechine Tavares, Marciane Magnani. & Evandro Leite de Souza. (2016). "Efficacy of the combined application of oregano and rosemary essential oils for the control of *Escherichia coli*, *Listeria monocytogenes* and *Salmonella Enteritidis* in leafy vegetables." *Food Control*, no. *59*, 468-477. doi: 10.1016/ j.foodcont.2015.06.017.

de Oliveira, Thales Leandro Coutinho, Gabriela de Barros Silva Haddad, Alcinéia de Lemos Souza Ramos, Eduardo Mendes Ramos, Roberta Hilsdorf Piccoli. & Marcelo Cristianini. (2019). "Optimization of high pressure processing to reduce the safety risk of low-salt ready-to-eat sliced turkey breast supplemented with carvacrol." *British Food Journal.*, doi: 10.1108/BFJ-10-2018-0646.

de Souza, Alana Gabrieli, Nathalie Mirelle Agostinho Dos Santos, Rondes Ferreira da Silva Torin. & Derval dos Santos Rosa. (2020). "Synergic antimicrobial properties of Carvacrol essential oil and montmorillonite in biodegradable starch films." *International Journal of Biological Macromolecules*, no. *164*, 1737-1747. doi: 10.1016/j.ijbiomac.2020.07.226.

de Souza, Evandro Leite. (2016). "The effects of sublethal doses of essential oils and their constituents on antimicrobial susceptibility and antibiotic resistance among food-related bacteria: A review." *Trends in Food Science & Technology*, no. *56*, 1-12. doi: 10.1016/j.tifs.2016.07.012.

de Souza, Evandro Leite, Geíza Alves de Azerêdo, Jossana Pereira de Sousa, Regina Celia Bressan Queiroz de Figueiredo. & Tânia Lúcia Montenegro Stamford. (2013). "Cytotoxic Effects of *Origanum vulgare* L. and *Rosmarinus officinalis* L. Essential Oils Alone and Combined at Sublethal Amounts on Pseudomonas fluorescens in a Vegetable Broth." *Journal of Food Safety*, no. *33* (2), 163-171. doi: 10.1111/jfs.12036.

Del Nobile, Matteo Alessandro, Annalisa Lucera, Cristina Costa, and Amalia Conte. (2012). "Food applications of natural antimicrobial compounds." *Frontiers in microbiology*, no. *3*, 287.

dos Santos Barbosa, Cristina Rodrigues, Jackelyne Roberta Scherf, Thiago Sampaio de Freitas, Irwin Rose Alencar de Menezes, Raimundo Luiz Silva Pereira, Joycy Francely Sampaio dos Santos, Sarah Silva Patrício de Jesus, Thais Pereira Lopes, Zildene de Sousa Silveira. & Cícera Datiane de Morais Oliveira-Tintino. (2021). "Effect of Carvacrol and Thymol on NorA efflux pump inhibition in multidrug-resistant (MDR) *Staphylococcus aureus* strains." *Journal of Bioenergetics and Biomembranes*, 1-10. doi: 10.1007/s10863-021-09906-3.

Falleh, Hanen, Mariem Ben Jemaa, Mariem Saada. & Riadh Ksouri. (2020). "Essential oils: A promising eco-friendly food preservative." *Food chemistry*, no. *330*, 127268. doi: 10.1016/j.foodchem.2020.127268.

Friedman, Mendel. (2015). "Antibiotic-resistant bacteria: prevalence in food and inactivation by food-compatible compounds and plant extracts." *Journal of agricultural and food chemistry*, no. *63* (15), 3805-3822. doi: 10.1021/acs.jafc.5b00778.

Gaysinsky, S., Davidson, P. M., Bruce, B. D. & Weiss, J. (2005). "Growth Inhibition of *Escherichia coli* O157:H7 and *Listeria monocytogenes* by Carvacrol and Eugenol Encapsulated in Surfactant Micelles." *Journal of Food Protection*, no. *68* (12), 2559-2566. doi: 10.4315/0362-028x-68.12.2559.

Gomes Neto, Nelson Justino, Marciane Magnani, Beatriz Chueca, Diego García-Gonzalo, Rafael Pagán. & Evandro Leite de Souza. (2015). "Influence of general stress-response alternative sigma factors σS (RpoS) and σB (SigB) on bacterial tolerance to the essential oils from *Origanum vulgare* L. and *Rosmarinus officinalis* L. and pulsed electric fields." *International Journal of Food Microbiology*, no. *211*, 32-37. doi: 10.1016/j.ijfoodmicro.2015.06.030.

Grompone, María A. (1991). "Propiedades físicas y químicas de las grasas bovinas fraccionadas e interesterificadas." *Grasas y aceites* no. *42*, 349-355.

Hague, M. A., Warren, K. E., Hunt, M. C., Kropf, D. H., Kastner, C. L., Stroda, S. L. & Johnson, D. E. (1994). "Endpoint temperature, internal cooked color, and expressible juice color relationships in ground beef patties." *Journal of Food Science*, no. *59* (3), 465-470. doi: 10.1111/j.1365-2621.1994.tb05539.x.

Hernández, Helga, Adéla Fraňková, Tomáš Sýkora, Pavel Klouček, Lenka Kouřimská, Iva Kučerová. & Jan Banout. (2017). "The effect of oregano essential oil on microbial load and sensory attributes of dried meat." *Journal of the Science of Food and Agriculture*, no. *97* (1), 82-87. doi: 10.1002/jsfa.7685.

Jaworska, Danuta, Elżbieta Rosiak, Eliza Kostyra, Katarzyna Jaszczyk, Monika Wroniszewska. & Wiesław Przybylski. (2021). "Effect of Herbal Addition on the Microbiological, Oxidative Stability and Sensory Quality of Minced Poultry Meat." *Foods*, no. *10* (7), 1537. doi: 10.3390/foods10071537.

Jørgensen, F., Sadler-Reeves, L., Shore, J., Heather Aird, Nicola Elviss, Fox, A., Kaye, M., Caroline Willis, Corinne Amar. & De Pinna, E. (2017). "An assessment of the microbiological quality of lightly cooked food (including sous-vide) at the point of consumption in England." *Epidemiology & Infection*, no. *145* (7), 1500-1509. doi: 10.1017/S0950268817000048.

Jouki, Mohammad. & Naimeh Khazaei. (2021). "Effects of active batter coatings enriched by quince seed gum and carvacrol microcapsules on oil uptake and quality loss of nugget during frying." *Journal of Food Science and Technology*, 1-10. doi: 10.1007/s13197-021-05114-4.

Juneja, Vijay K., Hari P Dwivedi. & Xianghe Yan. (2012). "Novel natural food antimicrobials." *Annual review of food science and technology*, no. *3*, 381-403. doi: 10.1146/annurev-food-022811-101241.

Juneja, Vijay K. & Mendel Friedman. (2008). "Carvacrol and cinnamaldehyde facilitate thermal destruction of *Escherichia coli* O157: H7 in raw ground beef." *Journal of food protection*, no. *71* (8), 1604-1611. doi: 10.4315/0362-028X-71.8.1604.

Juneja, Vijay K., Thippareddi, H. & Mendel Friedman. (2006). "Control of *Clostridium perfringens* in cooked ground beef by carvacrol, cinnamaldehyde, thymol, or oregano oil during chilling." *Journal of food protection*, no. *69* (7), 1546-1551. doi: 10.4315/0362-028X-69.7.1546.

Juneja, Vijay K., Ajit S Yadav, Cheng-An Hwang, Shiowshuh Sheen, Sudarsan Mukhopadhyay. & Mendel Friedman. (2012). "Kinetics of thermal destruction of *Salmonella* in ground chicken containing trans-cinnamaldehyde and carvacrol." *Journal of food protection*, no. *75* (2), 289-296. doi: 10.4315/0362-028X.JFP-11-307.

Juneja, V. K. & Eblen, B. S. (2000). "Heat inactivation of *Salmonella typhimurium* DT104 in beef as affected by fat content." *Letters in applied microbiology*, no. *30* (6), 461-467. doi: 10.1046/j.1472-765x.2000.00755.x.

Juneja, V. K., Eblen, B. S. & Marks, H. M. (2001). "Modeling non-linear survival curves to calculate thermal inactivation of *Salmonella* in poultry of different fat levels." *International journal of food microbiology*, no. *70* (1-2), 37-51. doi: 10.1016/S0168-1605(01)00518-9.

Kachur, Karina. & Zacharias Suntres. (2020). "The antibacterial properties of phenolic isomers, carvacrol and thymol." *Critical reviews in food science and nutrition*, no. *60* (18), 3042-3053. doi: 10.1080/10408398.2019.1675585.

Karam, Layal, Rayan Roustom, Mohamad G Abiad, Tahra El-Obeid. & Ioannis N Savvaidis. (2019). "Combined effects of thymol, carvacrol and packaging on the shelf-life of marinated chicken." *International journal of food microbiology*, no. *291*, 42-47. doi: 10.1016/j.ijfoodmicro.2018.11.008.

Kim, Nam Hee, Hye Won Kim, Hyeree Moon. & Min Suk Rhee. (2020). "Sodium chloride significantly enhances the bactericidal actions of carvacrol and thymol against the halotolerant species *Escherichia coli* O157: H7, *Listeria monocytogenes*, and *Staphylococcus aureus*." *LWT*, no. *122*, 109015. doi: 10.1016/j.lwt.2020.109015.

Lahou, Evy, Xiang Wang, Elien De Boeck, Elien Verguldt, Annemie Geeraerd, Frank Devlieghere. & Mieke Uyttendaele. (2015). "Effectiveness of inactivation of foodborne pathogens during simulated home pan frying of steak, hamburger or meat strips." *International journal of food microbiology*, no. *206*, 118-129. doi: 10.1016/j.ijfoodmicro.2015.04.014.

Li, Hui. & Michael Gänzle. (2016). "Effect of hydrostatic pressure and antimicrobials on survival of *Listeria monocytogenes* and enterohaemorrhagic *Escherichia coli* in beef." *Innovative Food Science & Emerging Technologies*, no. *38*, 321-327. doi: 10.1016/j.ifset.2016.05.003.

Lin, Chung-Tung Jordan. (2018). "Self-reported methods used to judge when a hamburger is ready at-home in a sample of US adults." *Food Control*, no. *91*, 181-184. doi: 10.1016/j.foodcont.2018.03.042.

Liu, Fang, Panpan Jin, Zhilan Sun, Lihui Du, Daoying Wang, Tong Zhao. & Michael P Doyle. (2021). "Carvacrol oil inhibits biofilm formation and exopolysaccharide production of *Enterobacter cloacae*." *Food Control*, no. *119*, 107473. doi: 10.1016/j.foodcont.2020.107473.

López-Pino, Jorge Isaac, Martin Valenzuela-Melendres, Juan Pedro Camou, Humberto González-Ríos, Fernando Ayala-Zavala. & Aida Peña-Ramos. (2021). "Predicción de la resistencia térmica de *Escherichia coli* O157: H7 en carne molida de res en función de la temperatura y las concentraciones de carvacrol y grasa." *Biotecnia*, no. *XXIII* (2), 8. doi: 10.18633/biotecnia.v23i2.1372 ["Prediction of thermal resistance of Escherichia coli O157:H7 in ground beef as a function of temperature and concentrations of carvacrol and fat." *Biotech*]

Lyon, B. G., Berry, B. W., Soderberg, D. & Nelson Clinch. (2000). "Visual color and doneness indicators and the incidence of premature brown color in beef patties cooked to four end point temperatures." *Journal of food protection*, no. *63* (10), 1389-1398. doi: 10.4315/0362-028X-63.10.1389.

Ma, Maomao, Junxin Zhao, Xianghui Yan, Zheling Zeng, Dongman Wan, Ping Yu, Jiaheng Xia, Guohua Zhang. & Deming Gong. (2022). "Synergistic effects of monocaprin and carvacrol against *Escherichia coli* O157: H7 and Salmonella Typhimurium in chicken meat preservation." *Food Control*, no. *132*, 108480. doi: 10.1016/j.foodcont.2021.108480.

Maldonado, Alma F., Andreas Schieber, & Michael G. Gänzle. (2015). "Plant defence mechanisms and enzymatic transformation products and their potential applications in food preservation: Advantages and limitations." *Trends in Food Science & Technology*, no. *46* (1), 49-59. doi: 10.1016/j.tifs.2015.07.013.

Marchese, Anna, Carla Renata Arciola, Erika Coppo, Ramona Barbieri, Davide Barreca, Salima Chebaibi, Eduardo Sobarzo-Sánchez, Seyed Fazel Nabavi, Seyed Mohammad Nabavi. & Maria Daglia. (2018). "The natural plant compound carvacrol as an antimicrobial and anti-biofilm agent: mechanisms, synergies and bio-inspired anti-infective materials." *Biofouling*, no. *34* (6), 630-656. doi: 10.1080/08927014.2018.1480756.

McIntyre, Lorraine, Virginia Jorgenson. & Mark Ritson. (2017). "Sous vide style cooking practices linked to *Salmonella Enteritidis* illnesses." *Environmental Health Review*, no. *60* (2), 42-49. doi: 10.5864/d2017-014.

Miladi, Hanene, Tarek Zmantar, Yassine Chaabouni, Kais Fedhila, Amina Bakhrouf, Kacem Mahdouani, & Kamel Chaieb. (2016). "Antibacterial and efflux pump inhibitors of thymol and carvacrol against food-borne pathogens." *Microbial Pathogenesis*, no. *99*, 95-100. doi: 10.1016/j.micpath.2016.08.008.

Montgomery, Douglas C. (2017). *Design and analysis of experiments*: John Wiley & sons.

Moon, Hyeree, Nam Hee Kim, Soon Han Kim, Younghoon Kim, Jee Hoon Ryu. & Min Suk Rhee. (2017). "Teriyaki sauce with carvacrol or thymol effectively controls *Escherichia coli* O157:H7, *Listeria monocytogenes*, *Salmonella Typhimurium*, and indigenous flora in marinated beef and marinade." *Meat Science*, no. *129*, 147-152. doi: 10.1016/j.meatsci.2017.03.001.

Munekata, Paulo ES, Mirian Pateiro, David Rodríguez-Lázaro, Rubén Domínguez, Jian Zhong. & Jose M Lorenzo. (2020). "The role of essential oils against pathogenic *Escherichia coli* in food products." *Microorganisms*, no. *8* (6), 924. doi: 10.3390/microorganisms8060924.

Nazzaro, Filomena, Florinda Fratianni, Laura De Martino, Raffaele Coppola. & Vincenzo De Feo. (2013). "Effect of essential oils on pathogenic bacteria." *Pharmaceuticals*, no. *6* (12), 1451-1474. doi: 10.3390/ph6121451.

Osaili, T. M., Griffis, C. L., Martin, E. M., Beard, B. L., Keener, A. E. & Marcy, J. A. (2007). "Thermal inactivation of *Escherichia coli* O157: H7, *Salmonella*, and *Listeria monocytogenes* in breaded pork patties." *Journal of food science*, no. *72* (2), M56-M61. doi: 10.1111/j.1750-3841.2006.00264.x.

Pateiro, Mirian, Rubén Domínguez, Roberto Bermúdez, Paulo ES Munekata, Wangang Zhang, Mohammed Gagaoua. & José M Lorenzo. (2019). "Antioxidant active packaging systems to extend the shelf life of sliced cooked ham." *Current Research in Food Science*, no. *1*, 24-30. doi: 10.1016/j.crfs.2019.10.002.

Pateiro, Mirian, Paulo ES Munekata, Anderson S Sant'Ana, Rubén Domínguez, David Rodríguez-Lázaro. & José M Lorenzo. (2021). "Application of essential oils as antimicrobial agents against spoilage and pathogenic microorganisms in meat products." *International journal of food microbiology*, no. *337*, 108966. doi: 10.1016/j.ijfoodmicro.2020.108966.

Picone, Gianfranco, Luca Laghi, Fausto Gardini, Rosalba Lanciotti, Lorenzo Siroli. & Francesco Capozzi. (2013). "Evaluation of the effect of carvacrol on the *Escherichia coli* 555 metabolome by using 1H-NMR spectroscopy." *Food Chemistry*, no. *141* (4), 4367-4374. doi: 10.1016/j.foodchem.2013.07.004.

Rinaldi, Massimiliano, Chiara Dall'Asta, Maria Paciulli, Martina Cirlini, Chiara Manzi. & Emma Chiavaro. (2014). "A novel time/temperature approach to sous vide cooking of beef muscle." *Food and bioprocess technology*, no. *7* (10), 2969-2977. doi: 10.1007/s11947-014-1268-z.

Rodriguez-Garcia, I., Silva-Espinoza, B. A., Ortega-Ramirez, L. A., Leyva, J. M., Siddiqui, M. W., Cruz-Valenzuela, M. R., Gonzalez-Aguilar, G. A. & Ayala-Zavala, J. F. (2016). "Oregano essential oil as an antimicrobial and antioxidant additive in food products." *Critical Reviews in Food Science and Nutrition*, no. *56* (10), 1717-1727. doi: 10.1080/10408398.2013.800832.

Sadek, Mustafa, Ahmed M Soliman, Hirofumi Nariya, Toshi Shimamoto, & Tadashi Shimamoto. (2021). "Genetic characterization of carbapenemase-producing *Enterobacter cloacae* complex and Pseudomonas aeruginosa of food of animal origin from Egypt." *Microbial Drug Resistance*, no. *27* (2), 196-203. doi: 10.1089/mdr.2019.0405.

Sarrazin, Sandra Layse F., Leomara A. da Silva, Ricardo B. Oliveira, Juliana Divina A. Raposo, Joyce Kelly R. da Silva, Fátima Regina G. Salimena, José

Guilherme S. Maia. & Rosa Helena V. Mourão. (2015). "Antibacterial action against food-borne microorganisms and antioxidant activity of carvacrol-rich oil from *Lippia origanoides* Kunth." *Lipids in Health and Disease*, no. *14* (1), 145. doi: 10.1186/s12944-015-0146-7.

Scandorieiro, Sara, Larissa C. de Camargo, Cesar A. C. Lancheros, Sueli F. Yamada-Ogatta, Celso V. Nakamura, Admilton G. de Oliveira, Célia G. T. J. Andrade, Nelson Duran, Gerson Nakazato. & Renata K. T. Kobayashi. (2016). "Synergistic and Additive Effect of Oregano Essential Oil and Biological Silver Nanoparticles against Multidrug-Resistant Bacterial Strains." *Frontiers in Microbiology*, no. *7*, 760. doi: 10.3389/fmicb.2016.00760.

Sharma, Shubham, Sandra Barkauskaite, Amit K Jaiswal, & Swarna Jaiswal. (2020). "Essential oils as additives in active food packaging." *Food Chemistry*, 128403. doi: 10.1016/j.foodchem.2020.128403.

Sheng, Haiqing, Yansong Xue, Wei Zhao, Carolyn J Hovde. & Scott A Minnich. (2020). "*Escherichia coli* O157: H7 curli fimbriae promotes biofilm formation, epithelial cell invasion, and persistence in cattle." *Microorganisms*, no. *8* (4), 580. doi: 10.3390/microorganisms8040580.

Šimat, Vida, Martina Čagalj, Danijela Skroza, Fausto Gardini, Giulia Tabanelli, Chiara Montanari, Abdo Hassoun. & Fatih Ozogul. (2021). "Sustainable sources for antioxidant and antimicrobial compounds used in meat and seafood products." In *Advances in food and nutrition research*, 55-118. Elsevier.

Siroli, Lorenzo, Giulia Baldi, Francesca Soglia, Danka Bukvicki, Francesca Patrignani, Massimiliano Petracci. & Rosalba Lanciotti. (2020). "Use of essential oils to increase the safety and the quality of marinated pork loin." *Foods*, no. *9* (8), 987. doi: 10.3390/foods9080987.

Stratakos, Alexandros Ch, Filip Sima, Patrick Ward, Mark Linton, Carmel Kelly, Laurette Pinkerton, Lavinia Stef, Ioan Pet. & Nicolae Corcionivoschi. (2018). "The *in vitro* effect of carvacrol, a food additive, on the pathogenicity of O157 and non-O157 Shiga-toxin producing *Escherichia coli*." *Food Control*, no. *84*, 290-296. doi: 10.1016/j.foodcont.2017.08.014.

Syed, Irshaan, Pratik Banerjee, and Preetam Sarkar. (2020). "Oil-in-water emulsions of geraniol and carvacrol improve the antibacterial activity of these compounds on raw goat meat surface during extended storage at 4°C." *Food Control*, no. *107*, 106757. doi: 10.1016/j.foodcont.2019.106757.

USDA. (2001). *Performance standards for production of certain meat and poultry products*. U. S. Food and Agriculture Department. Washington, D.C.: Office of Federal Register, National Archives and Records Administration.

Van Haute, S., Raes, K., Van der Meeren, P. & Sampers, I. (2016). "The effect of cinnamon, oregano and thyme essential oils in marinade on the microbial shelf life of fish and meat products." *Food Control*, no. *68*, 30-39. doi: 10.1016/j.foodcont.2016.03.025.

Vasan, Akhila, Renae Geier, Steve C Ingham. & Barbara H Ingham. (2014). "Thermal tolerance of O157 and non-O157 Shiga toxigenic strains of *Escherichia coli*, *Salmonella*, and potential pathogen surrogates, in frankfurter batter and ground beef of varying fat levels." *Journal of food protection*, no. *77* (9), 1501-1511. doi: 10.4315/0362-028X.JFP-14-106.

Villa, Tomás G. & Patricia Veiga-Crespo. (2013). *Antimicrobial compounds: current strategies and new alternatives*: Springer Science & Business Media.

Wang, Li, Jenneke Heising, Vincenzo Fogliano. & Matthijs Dekker. (2020). "Fat content and storage conditions are key factors on the partitioning and activity of carvacrol in antimicrobial packaging." *Food Packaging and Shelf Life*, no. *24*, 100500. doi: 10.1016/j.fpsl.2020.100500.

Wang, Yaying, Lifang Feng, Haixia Lu, Junli Zhu, Venkitanarayanan Kumar. & Xiaoxiang Liu. (2021). "Transcriptomic analysis of the food spoilers *Pseudomonas fluorescens* reveals the antibiofilm of carvacrol by interference with intracellular signaling processes." *Food Control*, no. *127*, 108115. doi: 10.1016/j.foodcont.2021.108115.

Wijesundara, Niluni M., Song F Lee, Zhenyu Cheng, Ross Davidson. & Vasantha Rupasinghe, H. P. (2021). "Carvacrol exhibits rapid bactericidal activity against *Streptococcus pyogenes* through cell membrane damage." *Scientific reports*, no. *11* (1), 1-14. doi: 10.1038/s41598-020-79713-0.

Zhuang, Zilin, Luchao Lv, Jiaxun Lu, Jinhang Lin. & Jian-Hua Liu. (2019). "Emergence of *Klebsiella pneumoniae* and *Enterobacter cloacae* producing OXA-48 carbapenemases from retail meats in China, 2018." *Journal of Antimicrobial Chemotherapy*, no. 74 (12):3632-3634. doi: 10.1093/jac/dkz394.

Chapter 3

Carvacrol as an Antibiofilm Agent in the Food Industry

**M. Melissa Gutiérrez-Pacheco[1],
Luis A. Ortega-Ramírez[1], A. Thalía Bernal-Mercado[2],
Cristóbal J. González-Pérez[3],
Samaria L. Gutiérrez-Pacheco[3]
and J. Fernando Ayala-Zavala[3],***

[1]Universidad Estatal de Sonora, San Luis Río Colorado, Sonora, Mexico
[2]Departamento de Investigacion y Posgrado en Alimentos, Universidad de Sonora, Hermosillo, Sonora, Mexico
[3]Centro de Investigación en Alimentación y Desarrollo, A. C., Hermosillo, Sonora, Mexico

Abstract

The ability of microorganisms to form biofilms is a growing concern in the food industry. Food processing areas include an environment that combines moisture and nutrients and becomes ideal for biofilm development. Biofilms are communities of microorganisms embedded in a self-produced matrix of extracellular polymeric substances, which protect against environmental stresses. The major problems associated with biofilms are their persistence and resistance to cleaning and sanitizing procedures, which represent high economic losses and public

---

* Corresponding Author's E-mail: jayala@ciad.mx.

In: A Closer Look at Carvacrol
Editor: Zak A. Cunningham
ISBN: 978-1-68507-627-6
© 2022 Nova Science Publishers, Inc.

health implications due to increased food spoilage and outbreaks. Although synthetic antimicrobials are approved in many countries, the trend has been the use of natural disinfectants, which fulfill the needs of today's consumers looking for safer, effective, and acceptable alternatives. Plant extracts such as essential oils represent a valuable source of biologically active molecules possessing antimicrobial properties. Specifically, carvacrol, the primary terpene compound of oregano essential oil, has shown high antimicrobial activity and inhibits biofilms of many food pathogens such as *Escherichia coli*, *Salmonella* Typhimurium, *Listeria monocytogenes*, *Campylobacter jejuni*, among others. Therefore, this chapter discusses the antibiofilm properties of carvacrol and their potential to control biofilm formation in meat, dairy, and fresh produce industries.

Keywords: natural disinfectant, plant-derived compound, food safety

Introduction

Food industry contamination by foodborne bacterial pathogens is an important health issue because it causes an increasing number of outbreaks worldwide every year (CDC, 2021). The World Health Organization (WHO) identifies food contamination as a global problem because consuming contaminated products causes diseases in millions (Hussain, 2016). Most food contamination cases are associated with microbial biofilms that can harbor and transmit spoilage and pathogenic bacteria, causing consecutive cross-contamination among different surfaces. Biofilms are microbial cells enclosed in a matrix of polymeric substances which protect the cells against environmental stresses such as temperature changes, pH, antimicrobials, among others (Giaouris & Simões, 2018a). For this reason, biofilms cells often remain on surfaces even after the regular cleaning and disinfection process. Biofilms can be formed on biotic and abiotic surfaces such as food and food contact surfaces, including tables, knives, conveyor belts, cutting boards, equipment, and floor, among others (Giaouris et al., 2014). Several factors promote biofilm development in the food industry, but nutrients and initial inoculum are the main factors. Different disinfectants are used in the food industry; however, some have little effect against bacterial cells living as biofilms.

Modern consumers are more aware of the toxic effects of synthetic disinfectants, demanding more natural and safer options. Essential oils (EOs) have been used for a long time for their antimicrobial properties. Specifically, carvacrol, a monoterpene found in the oregano EO, has a high spectrum of antimicrobial activity, inhibiting the biofilm formation of several Gram-negative and Gram-positive spoilage and pathogenic bacteria (Kachur & Suntres, 2020; Marinelli, Di Stefano, & Cacciatore, 2018). For example, Tapia-Rodriguez et al. (2017) showed a reduction in *Pseudomonas aeruginosa* biofilm formation at a concentration of 0.9–7.9 mM. Similarly, carvacrol reduced *E. coli* O157:H7 biofilm counts by 5.16 and 6.04 log CFU/cm^2, after being treated with 1% carvacrol for 30 s and 5 min, respectively, compared to controls (Campana & Baffone, 2018). Wang et al. (2021) reported inhibition of biofilm formation, exopolysaccharide production, and motility of the spoilage bacteria *Pseudomonas fluorescens* at 0.4 mM.

Several studies have reported that the antibiofilm activity of carvacrol is throughout several mechanisms, and this is dependent on the microorganism. In Gram-negative bacteria, carvacrol has inhibited the intercellular communication system called quorum sensing (QS), which controls biofilm formation in many bacteria. QS is characterized by the secretion and accumulation of signaling molecules called acyl-homoserine lactones (AHL), which in contact with the receptor protein, activates virulence genes related to biofilm formation, motility, and extracellular polymeric substances (EPS) production (Gutiérrez-Pacheco et al., 2019).

Carvacrol has inhibited AHL production in many bacteria, such as *P. aeruginosa* (Tapia-Rodriguez et al., 2019), *P. fluorescens* (Wang et al., 2021), *Pectobacterium carotovorum* (Joshi et al., 2016), among others. Tapia-Rodriguez et al. (2017) and Joshi et al. (2016) evidenced by molecular docking analysis that the antibiofilm activity of carvacrol was attributed to their interaction with the AHL synthase proteins. Carvacrol also acts at the genetic level, affecting the gene expression of QS proteins at subinhibitory concentrations in some bacteria (Burt, Ojo-Fakunle, Woertman, & Veldhuizen, 2014; Tapia-Rodriguez et al., 2019; Wang et al., 2021). Motility and EPS play important roles in biofilm formation, favoring bacterial adhesion and giving structure and protection to biofilm cells. Liu et al., (2021) reported that carvacrol inhibited motility and EPS synthesis in *Enterobacter cloacae*, which was correlated with the biofilm inhibition. In this sense, this chapter highlights the potential of carvacrol as an antibiofilm agent to be used in the food industry.

Biofilm Formation in the Food Industry

The World Health Organization (WHO) identifies food contamination as a global challenge because consuming contaminated food causes illness in millions of people (Hussain, 2016). Foodborne diseases are considered emergent public health problems around the world. The Centers for Disease Control and Prevention (CDC) reports (at July 2021) 9 multistate foodborne outbreaks by the consumption of cake mix (*E. coli* O121), prepackaged salads (*S.* Typhimurium), fully cooked chicken (*L. monocytogenes*), frozen cooked shrimp (*S.* Weltevreden), raw frozen breaded stuffed chicken products (*Salmonella* Enteritidis), cashew brie (*Salmonella* Duisburg), ground turkey (*Salmonella* Hadar), queso fresco (*L. monocytogenes*), and an unknown food source (*E. coli* O157:H7) (CDC, 2021). In general, the most common bacteria causing food contamination include *Campylobacter jejuni*, *E. coli*, *Staphylococcus aureus*, *Clostridium perfringens*, *Shigella* spp., *Vibrio parahaemolyticus*, *L. monocytogenes*, *Clostridium botulinum*, *V. vulnificus*, *Salmonella* spp., *Bacillus cereus*, *Vibrio cholerae*, and *Yersinia enterocolitica* (Hussain, 2016).

The presence of pathogenic microorganisms in foods results from its contamination in different stages of processing, such as handling, washing, cooking, cutting, packaging, storage, transportation, and others. Deficiencies in hygiene barriers, poor hygiene practices, and unhygienic equipment design promote food plant contamination (Muhterem-Uyar et al., 2015). However, even when cleaning and hygiene procedures are efficiently applied, microorganisms often remain and can firmly adhere to and form biofilms on every surface of food industry plants. These include stainless steel surfaces, conveyors belts, tables, floors, or equipment, making it difficult to remove with traditional disinfection procedures (Faille, Cunault, Dubois, & Benezech, 2018). Biofilm formation improves microbial survival to the environmental stresses of the food industry, such as refrigeration, acidity, salinity, disinfectants, among others.

Biofilms are aggregates of microbial cells embedded in a self-produced matrix of extracellular polymeric substances (EPS). Commonly, this is composed of polysaccharides, lipopolysaccharides, lipids, proteins, DNA, and other materials to stabilize the biofilm architecture and make cells more resistant to the disinfection processes (Gutiérrez-Pacheco et al., 2019). Biofilm formation involves five stages and can be developed in biotic and abiotic surfaces (Figure 1). Quorum Sensing (QS) is a bacterial communication system that relies on the secretion and detection of small diffusible signal molecules, allowing bacteria to synchronize in response to population density. QS regulates other virulence factors implicated in the biofilm formation process, such as motility, necessary to establish the adhesion surface. Biofilms may contain pathogenic and spoilage bacteria, resulting in contamination before and after processing, lower shelf-life of products, and increased numbers of foodborne illnesses (Shi & Zhu, 2009).

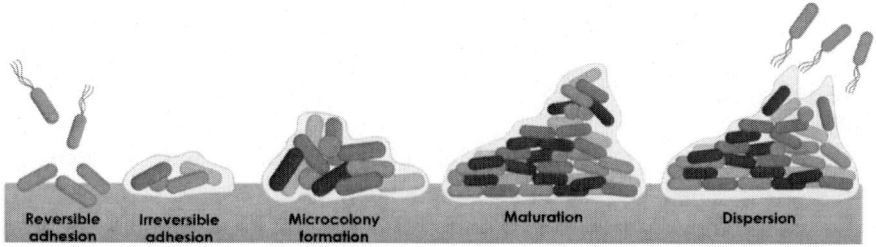

Figure 1. The biofilm formation process involves five stages: 1) reversible adhesion, 2) irreversible adhesion, 3) microcolony formation, 4) maturation, and 5) dispersion.

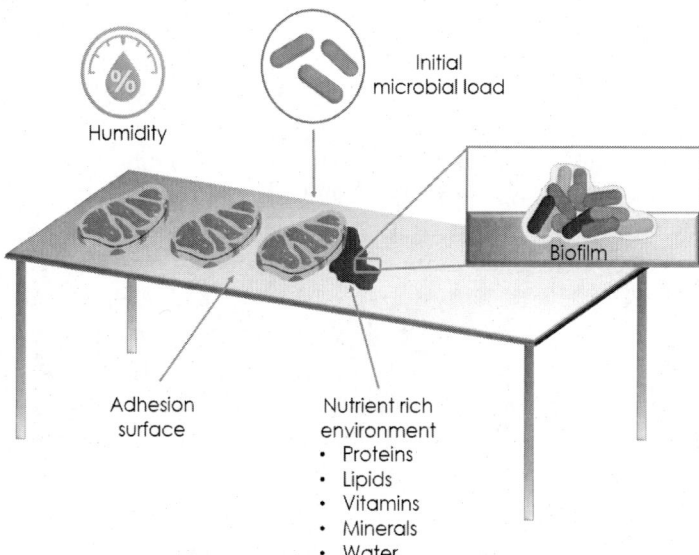

Figure 2. The food industry combines high humidity, a nutrient-rich environment, and an adhesion surface that promotes biofilm development.

The capacity of the different bacteria to form biofilms in food industries is facilitated by several factors (Figure 2), highlighting surfaces rich in nutrients and high humidity (Gutiérrez-Pacheco et al., 2019). Food-processing environments are rich in food residues, which create a conditioning film on industry surfaces, enhancing bacterial attachment. *Salmonella* spp., *E. coli*, and *L. monocytogenes* are biofilm-forming pathogens commonly found in the food industry (Coughlan, Cotter, Hill, & Alvarez-Ordóñez, 2016). Apart from pathogenic microorganisms, *Streptococcus*, *Enterococcus*, *Lactobacillus*, *Bacillus*, *Staphylococcus*, and *Pseudomonas* also have been isolated (Orhan-Yanıkan et al., 2019).

Fresh Produce Industry

In recent years, fresh produce has been recognized as an important vector for foodborne pathogens transmission and has increased every year worldwide. CDC reported from 2018 to date around 4200 cases of infected people in the United States after consuming fresh and processed fruits and vegetables contaminated with pathogens. The principal cases are salad mixes and fresh-cut fruits contaminated with *S.* Typhimurium, *E. coli* O157:H7, *Salmonella* Stanley, *S.* Enteritidis, *Salmonella* Newport, *Cyclospora*, *L. monocytogenes*, and *E. coli* O103 (CDC, 2018). Fresh produce is commonly associated with food outbreaks since they are food with high content of carbohydrates and high water activity, factors that promote microbial growth, and are consumed without a subsequent cooking process that reduces the microbial load.

The contamination is usually attributed to the nature of the practices carried out in the field, such as a bad implementation of good agricultural practices and poor hygiene during their subsequent processing (Giaouris & Simões, 2018a). Particularly in pre-harvest, pathogenic and spoilage microorganisms can come from water used for irrigation, organic fertilizers, animals in neighboring areas, or the direct contact of produce with the soil. Pre-harvest contamination may result in the introduction of pathogens into the processing plants, the biofilm development on food-contact surfaces, and cross-contamination, leading to an outbreak (Sela Saldinger & Manulis-Sasson, 2015).

Commonly, in the industry, fresh produce can become contaminated during the washing (due to improper disinfectant concentration and the presence of soil and dirt from previous washing), at the processing/packing stage, and during preparation due to lousy handling practices or insufficient sanitization of surfaces and equipment (Giaouris & Simões, 2018a). Rana et al. (2010) evaluated the cross-contamination of *Salmonella* during the postharvest washing of tomatoes. They evidenced that it spread from contaminated to uninoculated tomatoes during washing procedures. At least 20% of products are contaminated on the farm, while the rest of the outbreaks are by deficient hygiene practices after leaving the farm (Yaron & Römling, 2014).

Washing fresh produce with sanitizers is important in produce hygiene; it helps remove soil from produce and reduce the microbial load, avoiding future cross-contamination of sanitary products. However, washing is not completely effective in removing strongly adhered pathogenic bacteria on biofilms (Giaouris & Simões, 2018a). In addition, the surface characteristics of fresh produce, such as irregularities, roughness, crevices, and pits, favor bacteria adhesion, reducing the ability of washing or sanitizer treatments to remove or inactivate attached cells (Ukuku & Fett, 2006). Several authors have reported the influence of fresh produce surface on the ease of pathogens attachment. In particular, cantaloupe melon has a rough surface that allows microorganisms to attach to the texture to survive sanitizer treatments. Because of this, several outbreaks are associated with their consumption; in 2011, 147 people illness, and 33 died after consuming melon contaminated with *L. monocytogenes*. In 2012, there were 261 cases and three deaths from *S.* Typhimurium. Melon-associated outbreaks continued to occur in recent years (CDC, 2021).

Another factor influencing biofilm development in the fresh produce industry is the presence of nutrients. As mentioned before, food juices rich in nutrients form a conditioning layer that promotes bacteria attachment, resisting desiccation, and the action of antimicrobials. Kuda, Shibata, Takahashi, and Kimura (2015) evaluated the survival of *S. aureus*, *L. monocytogenes*, and *S.* Typhimurium on stainless steel surfaces before and after two hours of drying in the presence of distilled water and carrot juice. Results showed that bacterial counts were reduced in distilled water from 8 to 6, 5, and 3.4 logs CFU/dish for *S. aureus*, *L. monocytogenes*, and *S.* Typhimurium, respectively. In contrast, carrot juice protected the cells from desiccation, evidencing that proteins or carbohydrates in food contact surfaces increase the resistance of adherent bacteria. In a similar study, 50% aqueous solution of carrot and green bell pepper exerted a protective effect of adherent *S.* Typhimurium and *S. aureus* against hypochlorous acid (0.01%) and ethanol (70%) (Kuda, Koyanagi, Shibata, Takahashi, & Kimura, 2016).

In nature and food processing environments, many microorganisms exist attached to a surface as biofilms. For this reason, it is possible to assume that the incidence of outbreaks is due to biofilm-forming bacteria in pre- and post-harvest environments. Table 1 shows the most recent foodborne outbreaks in the United States of America associated with the consumption of fresh produce contaminated with biofilm-forming bacteria (CDC, 2021). Biofilms have been isolated from various fresh or processed fresh fruits and vegetables and industry surfaces. A study collected samples from fresh-cut industry surfaces such as peeler brushes, containers, conveyor belts, cutting boards, knife blades, slicer, knife handle, floor, and stainless steel tables. Among the 1000 tested isolates, major bacteria were biofilm-producers and identified as soil and plant-related bacteria, coliforms, and human-pathogenic bacteria (Liu et al., 2013).

Table 1. CDC report of USA foodborne outbreaks associated with the consumption of contaminated fresh produce in 2017-2021

Microorganism	Fresh produce	Cases/deaths	Year
Escherichia coli O157:H7	Leafy greens	25/1	2017
	Leafy greens	40/0	2020
	Romaine lettuce	65/0	2018
	Romaine lettuce	210/5	2018
	Romaine lettuce	167/0	2019
	Sunflower salad kit	10/0	2019
Escherichia coli O103	Clover sprouts	51/0	2020
Listeria monocytogenes	Enoki mushrooms	36/4	2020
Salmonella species	Maradol papayas	220/1	2017
	Bagged salad mix	701/0	2020
Salmonella Adelaide	Pre-cut melon	77/0	2018
Salmonella Montevideo	Raw sprouts	10/0	2018
Salmonella Javiana	Cut fruit	165/0	2018
Salmonella Uganda	Papaya	81/0	2019
Salmonella Carrau	Pre-cut melon	137/0	2019
Salmonella Stainley	Wood ear mushrooms	55/0	2020
Salmonella Enteritidis	Peaches	101/0	2020
Salmonella Newport	Onions	1127/0	2020
Salmonella Typhimurium	Prepackaged salads	11/0	2021

Warner, Rothwell, and Keevil (2008) reported high numbers of *Salmonella* Thompson on salad leaves (10^5 per square millimeter) aggregated as biofilms, which remained even after a washing process. Similarly, a total of 82 *Salmonella* isolates were recovered from fresh cabbage and spinach, and all were able to form biofilms (Isoken, 2015). Also, Han, Klu, and Chen (2017) reported the biofilm formation of *Salmonella* species and enterohemorrhagic *E. coli* on polystyrene, alfalfa, and bean sprouts. *S.* Enteritidis attached better to bean sprouts (5.58 log CFU/g) than alfalfa (5.28 log CFU/g), whereas *E. coli* formed more biofilm in polystyrene surfaces.

Another study showed the prevalence of *L. monocytogenes* in 512 packages of ready-to-eat vegetables marketed in São Paulo. Results showed *L. monocytogenes* in 3.1% of the samples with counts between 1.0×10^1 and 2.6×10^2 CFU/g. In addition, all isolates are also attached to stainless steel surfaces, reaching counts above 4 $\log_{10}$ CFU/cm^2 (Sant'Ana, Igarashi, Landgraf, Destro, & Franco, 2012). Delbeke, Ceuppens, Jacxsens, and Uyttendaele (2015) reported coliforms, *Salmonella* spp., and *E. coli* counts in fresh pre-packed basil and coriander leaves from a Belgian trading company.

Particularly, *S. enterica* and *E. coli* are one of the primary causal agents of fresh produce outbreaks. *Salmonella* is more frequent in fruits, sprouts, and vegetables such as cilantro, cauliflower, broccoli, lettuce, and spinach; *E. coli* is more frequent in leafy greens. Most *Salmonella*-infected people develop diarrhea, fever, and stomach cramps 6 hours to 6 days after being exposed to the bacteria. On the other hand, *E. coli* infections include severe diarrhea (often bloody), stomach cramps, vomiting, and in some cases, can cause the hemolytic uremic syndrome. Both bacterial infections can cause death in people without a correct treatment or immunosuppressed persons, children, and older people (CDC, 2021).

Meat Industry

Biofilms can be formed on many surfaces, being the processing surfaces of the meat industry one of the most common. Bacteria need environments rich in nutrients to proliferate mainly, proteins and carbohydrates, humidity, and good surfaces conditions (charge density, wettability, roughness, and stiffness, topography) (Zheng et al., 2021). In this sense, the food industry is rich in these components, making it susceptible to many contamination processes (Abebe, 2020). The biofilm formation in the meat industry is the primary source of contamination with foodborne pathogens that increase health risks. Corrosion and damaged equipment can also provoke economic losses (Lapointe, Deschênes, Ells, Bisaillon, & Savard, 2019).

The most prevalent microorganisms that attach to the meat processing surfaces and form biofilms are *L. monocytogenes*, *Salmonella* spp., enterohemorrhagic *E. coli*, *C. jejuni*, and *S. aureus*. Within these, *L. monocytogenes* is the most problematic microorganism in meat processing areas due to its capability to proliferate under refrigeration temperature, vacuum packaging, or in modified atmospheres in some meat products such as ready-to-eat foods (Giaouris & Simões, 2018b). In the USA, listeriosis caused a high mortality rate (15.6%) and was confirmed an increase of 1883 cases in 2013 to 2549 in 2018. This pathogen is responsible for 0.47 cases per 100,000 persons and is also a causal agent of hospitalizations and deaths in the USA (EFSA-ECDC, 2019; Silva et al., 2020).

Many studies have focused on evaluating the main surfaces and the most prevalent equipment contaminated with biofilms. The processing surfaces of the meat industry prone to be contaminated with biofilms are tables, knives, trays, cutters, stuffers, meat saws, grinders, floors, refrigerators, conveyor belts, and sinks, among others. This effect is because these materials are in contact with many organic samples such as meat, fat, blood, plasma, vitamins, and pigments that are the main constituents of the meat drip (Møretrø & Langsrud, 2004). Many of these surfaces have poor accessibility to good regular hygiene maintenance (Lindsay & Von Holy, 2006).

Wagner et al. (2020) evaluated 108 samples in a meat processing plant; 47 samples were taken before processing meat products, and 61 from other sites that were not in contact with foods. They identified with PCR analysis the presence of the species that conform to the biofilms and the different EPS formed. Among the sites evaluated, biofilms were detected in meat processing equipment such as cutter and screw conveyor. In contrast, for non-food contact areas, biofilms were detected in drains and water hoses. Biofilms samples from these surfaces were cultivated and identified mainly the genera *Brochothrix*, *Pseudomonas, Psychrobacter*, and other genera, indicating the formation of multi-species biofilms.

Bacterial communities in the processing line of a pork slaughterhouse were identified by massive sequencing 16S rRNA. The authors reported a wide variability of microorganisms in the different steps of animal processing. The main microorganisms identified in meat were the *Anoxybacillus*, *Chryseobacterium*, and *Moraxella* in singeing, polishing, evisceration, classification, and truck. On the other hand, *Anoxybacillus* and *Bacillus* were detected in polishing whips and gloves used in the evisceration and animal classification areas. The persistence of *Anoxybacillus* after heat treatments (scalding and singeing) is related to their biofilm formation capability (Zwirzitz et al., 2020).

Yin et al. (2018) evaluated the *Salmonella* biofilm formation on carcasses and polystyrene surfaces of processing lines of commercial beef plants at 25 and 37°C. There were isolated 77 strains and eight *Salmonella* serotypes such as *S.* Agona, *S.* Kingston, *S.* Kottbus, *S.* Calabar, *S.* Meleagridis, *S.* Senftenberg, *S.* Derby, and *S.* Typhimurium. Among these, *S.* Senftenberg and

S. Kingston were the higher biofilm producers at 25°C, whereas at 4°C, their biofilm formation capacity decreased.

In the meat industry, beef exudate, commonly named beef juice, is an important source of contamination because it is very prone to bacterial growth due to its nutrients composition. Ma et al. (2019) evaluated beef juice as a food-based model on 304-stainless steel coupons surfaces of the meat processing industry to evaluate the formation of biofilms of Shiga toxin-producing *E. coli* (STEC). Six STEC strains (O113, O145, O91, O157, O111, and O45) were grown with an M9 media or beef juice at 25, 50, and 100% to mimicking the nutrient profile present in meat processing areas as a source of nutrients. Also, the temperature effect (13 and 22°C) during 24, 48, and 72 h were evaluated. The effects observed in this study explained that beef juice promotes the formation of STEC's biofilms in a temperature, strain, and concentration-dependent manner. Besides, results showed that the low temperatures reduce biofilm formation capability, indicating that one of the most critical conditions in the meat industry is maintaining the cold chain through the processing line and efficient hygiene and sanitization practices (Ma et al., 2019).

Dairy Industry

The dairy industry is one of the most prevalent industries with contaminations associated with biofilms due to the nutrient-rich processing areas. A report by the CDC reported that 96% of foodborne diseases are associated with contaminated dairy foods. Many of these diseases have been related to unpasteurized milk and other products, causing 840 and 45 times more diseases and hospitalizations than the pasteurized milk or foods, respectively (Costard, Espejo, Groenendaal, & Zagmutt, 2017). Frequently, an increase in efforts to improve the cleaning and disinfection process has been made. The favorable conditions for bacterial growth and biofilm formation are proteins, fat, water, minerals, and peptides from milk, cheese, butter, and whey.

Biofilm formation in the dairy industry is frequently on milking containers, milk transport pipes, and accessories in the dairy industries (Srey, Jahid, & Ha, 2013). Also, processing areas with refrigeration temperatures and equipment made with stainless steel, allowing the growth and attachment of psychrotrophic bacteria. These microorganisms can contaminate raw or pasteurized milk and dairy products such as cheese, butter, and cream (Fusco et al., 2020). Previously, a study evaluated the effect of changing the surface properties of processing areas; particularly, a peptide-based coating was layered on stainless steel surfaces to determine their impact on bacterial adhesion. The coating significantly decreased the number of *Bacillus licheniformis,* and *P. aeruginosa* attached cells by around 2-log compared to the bare surfaces. Biofilm inhibition also was confirmed with confocal scanning laser microscopy. This study evidenced the impact of surface materials to promote biofilm development (Friedlander, Nir, Reches, & Shemesh, 2019).

Zou and Liu (2018) evaluated and identified biofilm-forming microorganisms in milk tanks, valves, pasteurizer pipes, heat-exchangers, pasteurizer tanks, filters, and homogenizers, using 16S rRNA gene sequence analysis. Sampled areas were grouped in pre-pasteurization, pasteurization, post-pasteurization, DSI (direct steam injection), and post-DSI. The identified strains were *Micrococcus* spp., *Clostridium* spp., *Acinetobacter* spp., *Enterobacter* spp., *Moraxella* spp., *Bacillus* spp., *Microbacterium oxydans, Serratia marcescens, Pseudomonas* spp., and *Staphylococcus epidermidis.* Then, biofilm production was evaluated on stainless steel and polystyrene surfaces at 37 and 55°C on different growth mediums. Among the strains, biofilms of *S. marcescens, Bacillus* spp., *Acinetobacter* spp., *Pseudomonas* spp., and *Clostridium* spp. were detected in the higher-temperature processing areas such as the pasteurizer heat-exchanger and the DSI-probes. Results indicated that the resistance of these strains to heat and the Cleaning-In-Place (CIP) process had been associated with their ability to form biofilms.

Lee et al. (2017) evaluated the ability of 85 strains of *L. monocytogenes* that were previously isolated from cheese, brine, and surfaces of two cheese processing plants to form biofilms on polystyrene and stainless steel surfaces. Twenty-one isolates were detected as biofilm producers in both surfaces, Karaca, Buzrul, and Coleri Cihan (2019) assessed the capacity of several *Anoxybacillus, Geobacillus,* and reference strains in milk to form biofilms in abiotic surfaces. These surfaces were glass, stainless steel, polyvinyl chloride (PVC), and polypropylene at 55 and 65°C. Also, the effect of the type of milk (skim, semi-skim, whole) was evaluated. The highest cell counts (>4 log CFU/cm^2) were observed for *G. thermodenitrificans* on glass and stainless steel at 55 and 65 °C, respectively. Moreover, the milk used directly influenced bacteria, being *Anoxybacillus* members the major biofilm-formers in skim milk than in semi-skim and whole milk.

Additionally, it was found that *Geobacillus* biofilms increase as the fat content in the milk increase. These findings were related to the effect of butyric acid on genes responsible for biofilm formation, such as *tapA gene*.

Biofilm formation by spoilage and pathogenic bacteria in the fresh produce, meat, and dairy industry is a global contamination issue, existing and emerging at all times. In this sense, new approaches are needed to reduce the contamination in the food industry, which contemplate biofilm formation inhibition. Among the alternatives, carvacrol has shown high antimicrobial and antibiofilm activity. For this reason, it can be used as a natural antimicrobial agent to inhibit biofilm formation in the food industry and offset the limitations of conventional disinfectants (Giaouris et al., 2014).

Sources and Extraction of Carvacrol

Carvacrol is an oxygenated phenolic monoterpene with methyl and isopropyl substitutions. Figure 3 shows the structural formula and main physical and chemical properties of carvacrol. Carvacrol ($C_{10}H_{14}O$) is also chemically known as 2-methyl-5-(1-methyl ethyl)-phenol and with some synonyms: 5-isopropyl-2-methyl phenol, isopropyl-o-cresol, p-cymene-2-ol, 2-hydroxy-p-cymene, and iso-thymol (De Vincenzi, Stammati, De Vincenzi, & Silano, 2004).

Carvacrol is a natural constituent of essential oil (EO) of aerial organs of some aromatic plants. The most significant sources of carvacrol are oregano, thyme, and savory herbs traditionally used as folk remedies and food seasoning due to their biological activities, great flavor, and aroma. Several plants of the *Lamiaceae* family are rich sources of carvacrol, such as oregano (*Origanum vulgare, O. majorana, O. ehnrebergii, O. syriacum, O. dictmanuus, O. scabrum, O. onites)*, thyme (*Thymus vulgaris, T. glandulosus, T. schimperi, T. zygis*), savory *(Satureja hortensis, S. montana, S. thymbra, S. cuneifolia)*, bee balm *(Monarda didyma, M. fistulosa)*, Thymbra *(Thymbra spiccata, T. calostachya, T. capitata),* and *Zataria multiflora* (Can Baser, 2008; Côté, Pichette, St-Gelais, & Legault, 2021; Krisilia, Deli, Koutsaviti, & Tzakou, 2021; Marinelli et al., 2018; Rota, Herrera, Martínez, Sotomayor, & Jordán, 2008; Saedi Dezaki et al., 2016; Zgheib et al., 2019).

Other species such as *Lippia graveolens* plant, *Nigella sativa* seeds, among others from different families, present good carvacrol content (Ghahramanloo et al., 2017; Rodriguez-Garcia et al., 2016).

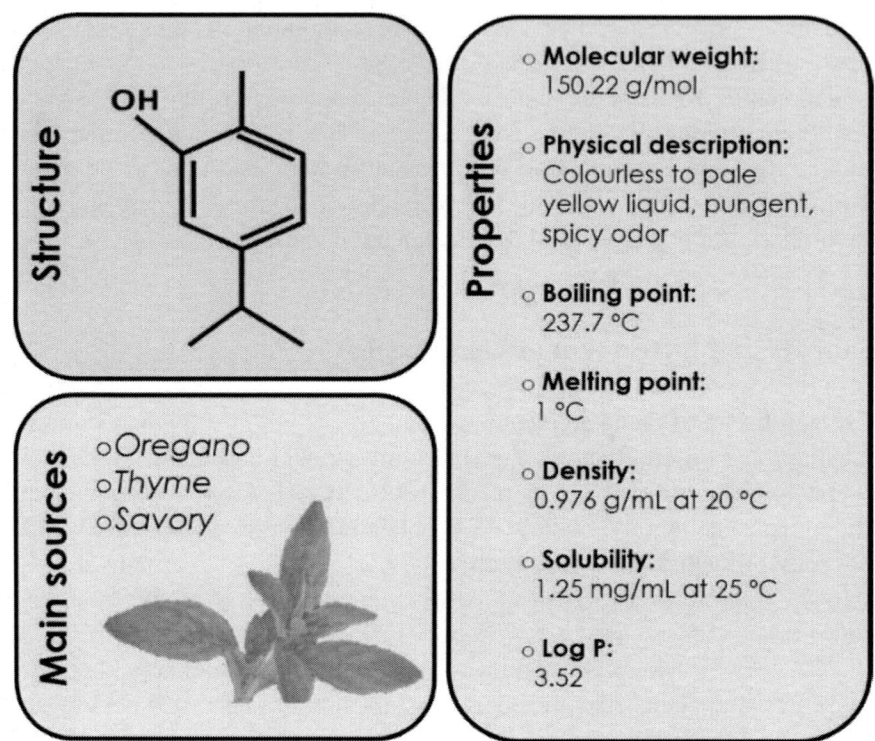

Figure 3. Carvacrol structure and main physical and chemical properties.

Depending on plant genetic, climatic factors, seasonal variability, geographic zones, collection period, plant organs, and extraction techniques, the biosynthesis and proportion of carvacrol in EOs may variate, ranging carvacrol from 3 – 87% in EOs of different and same species of aromatic plants (Table 1) (Aghaei, Hossein Mirjalili, & Nazeri, 2013; Mechergui et al., 2010; Morshedloo, Salami, Nazeri, Maggi, & Craker, 2018; Novak, Lukas, & Franz, 2010; Pourhosseini, Ahadi, Aliahmadi, & Mirjalili, 2020).

For example, Morshedloo et al. (2018) observed a broad variability among the EO composition of seven populations of Iranian oregano from different bioclimate, geographical zones, and harvest years but under the same soil and climate conditions. Specifically, carvacrol variation ranged from 0.36% to 46.86%.

Another study has revealed a wide variation in the qualitative and quantitative composition and yield from *O. vulgare* EO depending on the harvest year; for example, carvacrol content ranged from one year to another from 1.7 to 2.7% (Mechergui et al., 2010). Similarly, Pourhosseini et al. (2020) reported a variation from 28.3 to 95.2% in carvacrol content of EOs from seven *Zataria multiflora* populations of Iran. Therefore, it is essential to perform a phytochemical and biological characterization of EOs before possible industrial applications since its carvacrol content, and biological activity may vary naturally from different sources.

Carvacrol can be extracted from the EO of several plants by a variety of extraction methods. Conventional procedures, including steam distillation, hydrodistillation, and non-polar solvent extractions, are the most frequent method to obtain carvacrol-rich EOs because they are simple, cheap, and well-accepted. Nevertheless, conventional procedures also have low efficiency, long extraction time, large amounts of solvents, and toxic solvent residues (Sahraoui, Hazzit, & Boutekedjiret, 2017). As an alternative, some techniques have been developed to overcome these problems as new clean, affordable, and highly efficient technology with the obtention of a high-quality product (Tongnuanchan & Benjakul, 2014). Among them, we can find solvent-free microwave extraction, supercritical fluids extraction, ultrasound-assisted extraction, and pulsed electrical-assisted extraction, which have been gaining interest in obtaining essential oil from different plant sources. The isolation and purification of carvacrol may be conducted after extracting plant EO by remotion of others compounds and using different organic solvents and chromatographic techniques (column, thin-layer) to fractionate and separate the molecule (Galehassadi, Rezaii, Najavand, Mahkam, & Mohammadzadeh, 2014).

Table 1. Occurrence of carvacrol in plants essential oils

Essential oil source	Plant material	Collection site	Carvacrol content (%)	Extraction method	Reference
Lippia graveolens	Dried leaves	Mexico	30.89 – 36.24	Steam distillation	Soto-Armenta et al. (2017)
Lippia graveolens	Dried leaves	Mexico	6.75	Solvent extraction	González-Trujano et al. (2017)
Monarda didyma	Aerial parts	Canada	49.03	Steam distillation	Côté et al. (2021)
Monarda fistulosa	Air-dried flowers and leaves	United States of America	45.7 – 71.5	Hydrodistillation	Ghosh et al. (2020)
Nigella sativa	Seeds	Iran	10	Hydrodistillation	Kazemi (2015)
Nigella sativa	Seeds	Malaysia	1.82	Supercritical fluid extraction and cold press	Mohammed et al. (2016)
Nigella sativa	Seed oil	India, Iran	1.65 – 4.83	Supercritical fluid extraction and solvent extraction	Ghahramanloo et al. (2017)
Origanum ehrenbergii	Dried aerial part	Lebanon	48.1-88.6	Hydrodistillation	Zgheib et al. (2019)
Origanum mejorana	Dried and fresh branch, leaf, and herb	Turkey	56.40-86.47	Hydrodistillation	Bagcı, Kan, Dogu, and Çelik (2017)
Origanum minutiflorum	Aerial parts	Turkey	84.4	Hydrodistillation	Baydar, Sağdiç, Özkan, and Karadoğan (2004)
Origanum onites	Aerial parts	Turkey	86.9	Hydrodistillation	Baydar et al. (2004)
Origanum syriacum	Dried herb	-	82.8 - 86.8	Ultrasonic extraction	Novak et al. (2010)
Origanum syriacum	Entire plant	Egypt	81.38	Hydrodistillation	El Gendy et al. (2015)
Origanum vulgare	Dried leaves and inflorescences	Iran	0.36 – 46.86	Hydrodistillation	Morshedloo et al. (2018)
Origanum vulgare	Dried herb	Canada	23.8 – 27.9	Ultrasonic extraction	Novak et al. (2010)
Origanum vulgare	Aerial parts	India	6.90	Hydrodistillation	Raina and Negi (2012)
Satureja cuneifolia	Aerial parts	Turkey	53.3	Hydrodistillation	Baydar et al. (2004)

Essential oil source	Plant material	Collection site	Carvacrol content (%)	Extraction method	Reference
Satureja hortensis	Dried aerial parts	Turkey	67	Hydrodistillation	Mihajilov-Krstev et al. (2009)
Satureja hortensis	Dried aerial parts	Romania	13.22	Hydrodistillation	Popovici et al. (2019)
Satureja thymbra	Dried aerial part	Greece	65.2	Hydrodistillation	Krisilia et al. (2021)
Thymbra calostachya	Dried aerial part	Greece	73.6 – 78.4	Hydrodistillation	Krisilia et al. (2021)
Thymbra capitata	Dried aerial part	Greece	50.4 – 78.1	Hydrodistillation	Krisilia et al. (2021)
Thymbra spicata	Aerial parts	Turkey	75.5	Hydrodistillation	Baydar et al. (2004)
Thymus hyemalis	Dried aerial parts	Spain	40.1	Steam distillation	Rota et al. (2008)
Thymus lanceolatus	Air-dried aerial parts	Algeria	3.57	Hydrodistillation	Khadir et al. (2016)
Thymus vulgaris	Leaves	Brazil	45.5	Steam distillation	Fachini-Queiroz et al. (2012)
Thymus vulgaris	Dried aerial parts	Spain	2.8	Steam distillation	Rota et al. (2008)
Thymus zygis	Dried aerial parts	Spain	3.5	Steam distillation	Rota et al. (2008)
Zataria multiflora	Dried aerial part	Iran	28.85	Percolation method	Saedi Dezaki et al. (2016)

Antimicrobial and Antibiofilm Properties of Carvacrol Against Food Pathogens

Antimicrobial

Carvacrol is a natural phenolic compound recognized for its high antimicrobial activity against deteriorative and pathogenic bacteria and fungi. (Cacciatore et al., 2020). The use of carvacrol as an antimicrobial compound has been exploited in different areas such as food hygiene, cosmetics, and medicine (Raut & Karuppayil, 2014). The broad spectrum of antimicrobial activity could be used in the food industry because carvacrol is generally recognized as safe (GRAS). For this reason, several studies have shown their potential use as a natural disinfectant in the food industry (Sharifi-Rad et al., 2018).

Carvacrol has inhibited spoilage and pathogenic bacteria implicated in contamination of fresh produce, dairy, meat, and fish food industry (Table 3). Among the inhibited pathogenic bacteria are *Salmonella, E. coli, L. monocytogenes, Y. enterocolitica, C. perfringens, C. jejuni,* and *S. aureu*s. Stratakos et al. (2018) reported a potent antibacterial effect of carvacrol against eight Shiga toxin-producer *E. coli* strains, showing a MIC of 0.031%. At this concentration, carvacrol damaged the cell membrane, increasing permeability and reducing intracellular ATP levels. Additionally, sub-inhibitory concentrations (1/2 MIC) decreased the adherence of *E. coli* O157:H7 to intestinal cells.

The growth of *Acitenobacter baumanni* and *E. coli* isolated from the meat industry were significantly reduced by carvacrol at concentrations between 0.15 and 0.25 mg/mL, showing reductions of approximately 8 and 9 log CFU/mL, respectively (Orhan-Yanıkan et al., 2019). Takahashi et al. (2021) reported that carvacrol suppressed spoilage lactic acid bacteria (which causes problems in salad dressings, tomato-based products, and processed meats) at 0.5 mg/mL. In addition, the counts of inoculated sausages were lower at 5 mg/g and completely suppressed at 20 mg/g.

Table 2. Antimicrobial activity of carvacrol against food-contamination microorganisms

Microorganism	Common contaminated food	Dose of inhibition	References
Pseudomonas fluorescens	Fish, poultry, dairy, and other fresh products	MIC: 1.6 mM	(Wang et al., 2021)
Pseudomonas fluorescens ΔluxI mutant		MIC: 1.4 mM	
Listeria monocytogenes	Ready-to-eat food	MIC: 0.06% MBC: 0.09%	(Ashrafudoulla, Rahaman Mizan, Park, & Ha, 2021)
Pseudomonas aeruginosa	Aquatic products	MIC: 0.03% MBC: 0.05%	
Pseudomonas fluorescens	Fish, poultry, dairy, and other fresh products	MIC: 93 mg/L MBC:117 mg/L	(Mauriello, Ferrari, & Donsì, 2021)
Staphylococcus epidermidis	Meat and dairy products	MIC:117 mg/L MBC:298 mg/L	
Lactobacillus carnosum	Ham, sausage, and other meat products	1000 mg/L	(Takahashi et al., 2021)
Lactobacillus citreum	Vegetables, fermented sausages, and cheeses		
Lactobacillus plantarum			
Enterococcus faecalis			
Escherichia coli	Vegetables and meat	MIC: 0.32 mg/mL	(Rao et al., 2021)
Staphylococcus aureus			
Bacillus cereus	Instant foods	MIC: 0.3875 mg/mL	(Rao et al., 2020)
Salmonella	Meat, poultry, and eggs	MBC: 0.775 mg/mL	
Salmonella Enteritidis	Egg and poultry	MSC: 120 μg/mL	(Barbosa et al., 2020)
Staphylococcus aureus	Aquatic products	MIC: 1.6 mM	(Wang, Hong, Liu, Zhu, & Chen, 2020)
Pseudomonas fluorescens	Fish, poultry, dairy, and other fresh products	MIC: 2.4 mM	
Escherichia coli	Vegetables and meat	MIC: 0.70 mg/mL	(Requena, Vargas, & Chiralt, 2019)
Listeria innocua	Dairy products	MIC: 0.75 mg/mL	
Acinetobacter baumannii	Meat	MBC: 0.1 mg/mL	(Orhan-Yanıkan et al., 2019).

Table 2. (Continued)

Microorganism	Common contaminated food	Dose of inhibition	References
Escherichia coli	Vegetables and meat	MBC: 0.2 mg/mL	(Sokolik, Ben-Shabat-Binyamini, Gedanken, & Lellouche, 2018)
Pectobacterium carotovorum	Vegetables	MIC: 2.66 mM MBC 3.99 mM	(Gutierrez-Pacheco et al., 2018)
Escherichia coli O157:H7	Vegetables and meat	MBC: 0.08 mg/mL	(García-Heredia, García, Merino-Mascorro, Feng, & Heredia, 2016)
Penicillium digitatum	Citrus fruits	MIC: 0.125 mg/mL MFC: 0.25 mg/mL	(Yang et al., 2021)
Botrytis cinerea	Fruits and vegetables	MIC: 7.81 mg/L MFC: 15.63 mg/L	(Zhao et al., 2021)

MIC: Minimal inhibitory concentration, MBC: Minimal bactericidal concentration, MFC: Minimal fungicidal concentration, MSC: Maximum sublethal concentration.

The antibacterial mode of action of carvacrol is related to its hydrophobicity. Carvacrol has a partition coefficient (LogP) of 3.52, which indicates their affinity to interact with the bacterial membrane, causing its destabilization, increasing fluidity and permeability, which release the cytoplasmic content and cell death. In addition, their hydroxyl group on the aromatic ring and the electron delocalization in the aromatic ring (double bonds) makes carvacrol a proton exchanger, causing an alteration of the gradient across the membrane, the collapse of the proton motive force, and a depletion of the ATP pool (Marchese et al., 2018).

The antimicrobial activity of carvacrol also was tested against the natural yeasts of grapes and those that cause wine spoilage. Carvacrol (64 mg/mL) inhibited spoilage yeast growth in wine, showing better activity than the commercially used wine preservative potassium metabisulphite. Similarly, the antifungal activity of carvacrol against *B. cinerea* was determined by Zhao et al. (2021). Results showed significant inhibition of spore germination and the mycelial growth of *B. cinerea*. Furthermore, in the *in vivo* tests, the antifungal activity of carvacrol in the vapor phase (125 mg/L) suppressed the decomposition of cherry tomatoes by 96.39% compared to the control fruit. The mechanism by which carvacrol exerted its antifungal action was the inhibition of ergosterol synthesis, membrane damage, and cytoplasmic content leakage (Chavan & Tupe, 2014).

Antibiofilm

Bacterial persistence and resistance to disinfection processes are attributed to their biofilm formation ability on food and food contact surfaces. Carvacrol has inhibited the biofilm formation of many spoilage bacteria like *Staphylococcus*, *Lactobacillus*, *Bacillus*, and *Pseudomonas;* this effect was also observed against pathogenic Escherichia *Campylo-bacter*, *Clostridium*, *Listeria*, and *Salmonella*, implicated in food outbreaks (CDC, 2021). Carvacrol significantly reduced *L. monocytogenes* and *P. aeruginosa* biofilms on an MBEC™ biofilm device by 3.81 and 5.04 log CFU/peg at 0.12 and 0.06% carvacrol, respectively. On the polypropylene surface, reductions were 4.62 and 4.79 log CFU/cm^2, respectively (Ashrafudoulla, Rahaman Mizan, et al., 2021).

Espina, Berdejo, Alfonso, García-Gonzalo, and Pagán (2017) evaluated the disinfectant potential of carvacrol (500–2000 μL/L) against mature biofilms of *L. monocytogenes*, *S. aureus*, or *E. coli*. Carvacrol reduced biofilm cells of the three bacteria for more than 5 log cycles at 1000 ppm at 45°C for 60 min.

Carvacrol at 2.5 µL/mL also showed antibiofilm activity against *S. aureus* young and mature biofilms grown on stainless steel surfaces, showing reductions of 3.5 log/cm^2. This effect was similar to those exerted by sodium hypochlorite at 250 mg/L. In addition, microscopy analysis evidenced the membrane damage caused by carvacrol on biofilm cells (Rodrigues et al., 2018).

Carvacrol solution (1% v/v) was tested against *E. coli* O157:H7 and *S. aureus* biofilms developed for four days on a stainless steel surface. Results showed 5.16 and 6.04 log CFU/cm^2 reduction of *E. coli* O157:H7 ATCC 35150 biofilms after carvacrol treatment for 30 s and 5 min, respectively, compared to control; in contrast, at 15 min of contact, no colonies were detected. A similar trend was observed in *S. aureus* biofilms, reducing 4.45 and 5.66 log CFU/cm^2 after 30 s and 5 min of contact with carvacrol (Campana & Baffone, 2018).

It has been reported that multi-species biofilms are more resistant to be eliminated. This effect was evidenced by Wang et al. (2020), which evaluated the effect of carvacrol (0.4, 0.8, and 1.6 mM) against *S. aureus* and *P. fluorescens* mono and dual biofilms formed on stainless steel coupons in fish juice. *S. aureus* cell counts in mono- and dual-species biofilm were reduced by 2.2 and 1.6 log CFU/cm^2 for 48 h after 1.6 mM carvacrol treatment, respectively. Similarly, *P. fluorescens* log reductions were 1.4 and 0.8 log CFU/cm^2 for mono- and dual-species biofilm, respectively. Additionally, the exopolysaccharide secretion was reduced in a dose-dependent manner without affecting cell viability, especially for *S. aureus*.

The discovery that the QS system regulates biofilm formation in many bacteria has made recent studies inhibiting these communication systems. QS systems can be distinguished: the acyl-homoserine lactone molecules (AHL) in Gram-negative bacteria, the autoinducing peptide (AIP) in Gram-positive bacteria, and the autoinducer-2 (AI-2) in both Gram-positive and Gram-negative bacteria (Brackman & Coenye, 2015). In the AHL-QS system of Gram-negative bacteria, once the concentration of AHL increases to threshold levels, an active complex is formed by the binding of AHL to LuxR receptor proteins, which has a high affinity to specific DNA sequences; called "lux boxes." These sequences are in promoter regions of genes in the QS regulon. They activate their respective virulence factors such as biofilm formation, motility, and EPS synthesis, which helps their development. Some bacterium uses flagella-mediated motility to overcome the repulsion forces and reach the adhesion surface. On the other hand, EPS synthesis allows bacteria to attach strongly to surfaces and create the tridimensional structure of biofilm, which helps resist antimicrobial action.

Several studies evidenced the effect of carvacrol against the QS system; even it has been designated as a potential QS inhibitor. The LuxI/LuxR mediated QS signaling regulates the strong biofilm-forming ability of *P. fluorescens*, increasing their resistance to the disinfection process and persistence in the fresh and cold food chain. An investigation by Wang et al. (2021) explored the effect of carvacrol in the biofilm formation *P. fluorescens* strains (PFuk4 and PF07) and their luxI (ΔluxI) mutants. These authors found that 0.4 mM of carvacrol decreased the C4-HSL and C6-HSL content, molecules of the QS system, in the WT strains. This effect was reflected in a lower biofilm formation, exopolysaccharide production, and motility of *P. fluorescens*. Tapia-Rodriguez et al. (2019) reported that carvacrol (1.9 mM) reduced *lasR* expression in *P. aeruginosa* without affecting *lasI*. This result evidenced the possible affectation of LasI activity, the AHL-producer protein. Furthermore, it was corroborated by the 60% lower AHLs compared to controls without affecting cellular viability. The lower AHL synthesis was reflected in lower biofilm formation and swarming motility.

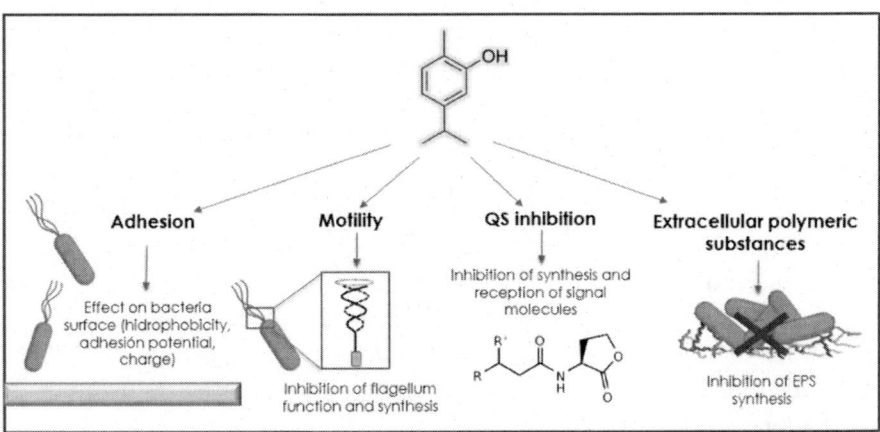

Figure 4. Proposed mechanism of action of carvacrol against food spoilage and pathogenic bacteria.

On the other hand, *E. cloacae* biofilm formation was significantly reduced at 64 and 128 μg/mL of carvacrol during cultivation for 72 h. Also, the EPS content in *E. cloacae* biofilms markedly decreased at 12, 24, 48, and 72 h cultivation, compared with the control samples. Additionally, the transcriptional analysis showed a down-regulation of 24 of 29 selected genes, such as those of curli fimbriae (*csgABCEFG*) and colonic acid polysaccharide genes (*wcaABCDEFKLM, wza, wzb, wzc, gmd, manB, manC*), which are important for biofilm formation (Liu et al., 2021). On the other hand, carvacrol inhibited and eradicated the biofilm development of the soft rot pathogen *P. carotovorum* on polypropylene surfaces at 1.33 and 3.99 mM, respectively. In addition, a sublethal concentration of carvacrol (0.66 mM) reduced EPS and swimming motility compared to the untreated bacteria. Furthermore, polysaccharides were the main components of the *P. carotovorum* biofilm matrix, whose synthesis was inhibited by the presence of carvacrol (Gutierrez-Pacheco et al., 2018).

Application of Carvacrol in the Food Industries

Carvacrol has shown effective activity against biofilm formation by various bacteria when applied directly to vinegar, fruit juices, and minced meats (Friedman, 2014). The direct application of carvacrol in food or food surfaces has been examined and accepted by international authorities. This component of oregano has been classified as GRAS for Food and Drug Administration (FDA), so it could be used in food for human consumption. Its safety, sensory attributes, and antibacterial and antibiofilm activity make this compound a good choice for the food industry.

The application of carvacrol on different surfaces is an excellent alternative to combat biofilm formation. Different investigations demonstrated antibiofilm activity of carvacrol in stainless steel, polypropylene, and polystyrene surfaces against Gram-positive bacteria, like *L. monocytogenes* and *S. aureus*, and Gram-negative bacteria like *E. coli* and *P. aeruginosa* (Ashrafudoulla, Mizan, Park, & Ha, 2021; Gómez-Sequeda, Cáceres, Stashenko, Hidalgo, & Ortiz, 2020).

Tapia-Rodriguez et al. (2017) reported that carvacrol reduces up to 70% of *P. aeruginosa* biomass production in stainless steel surface; therefore, the biofilm inhibition is a fact since they verified the correlation between biomass production and bacteria embedded within the biofilm. Other surfaces in food industries that are susceptible to biofilm formation are polyvinylchloride, polypropylene, and polyethylene. These surfaces were proved for Walczak, Michalska-Sionkowska, Olkiewicz, Tarnawska, and Warżyńska (2021), who showed that carvacrol inhibited biofilm of *S. aureus* up to 88% and *P. aeruginosa* up to 100%. These results showed that carvacrol could be used in the food industry's equipment, transport, or storage surfaces.

Carvacrol and other EOs possess a high aroma and volatility, low stability, and susceptibility to environmental changes that could affect their efficient application in the food industry (Cacciatore et al., 2020). For this reason, some studies have encapsulated carvacrol to maintain or improve their antibacterial and antibiofilm activity. In addition, considering the high moisture content in the food industry, this could promote a controlled release of this compound in the food industry environment. The forms of application of carvacrol in the food industry include encapsulation, incorporation in a coating film, into packaging materials, as nanoemulsions, and combination with other antibacterial agents, among others (Kachur & Suntres, 2020; Mittal, Rana, & Jaitak, 2019; Nostro & Papalia, 2012).

Incorporating carvacrol in different coatings is an alternative to the food industry as antibacterial or antibiofilm packaging. Jahdkaran, Hosseini, Mohammadi Nafchi, and Nouri (2021), show that carvacrol incorporated in methylcellulose coatings in polyethylene films has antibacterial activity against *L. innocua, E. coli,* and *S. aureus*. Several investigations have shown that encapsulated carvacrol maintains its biological characteristics and may have a long-term release.

Cacciatore et al. (2020) encapsulated carvacrol in nanoliposomes and nanocapsules with a preformed polymer and evaluated their effect against *L. monocytogenes, S. aureus, E. coli,* and *Salmonella* spp. adhered to stainless steel. *Salmonella* and *E. coli* counts were reduced below the detection limit using the carvacrol nanocapsules at 4.42 mg/mL. Although free carvacrol has better results than encapsulated, nanocapsules have controlled carvacrol release and masked its aroma, so it becomes an interesting option to disinfect food surfaces since it would maintain its activity for long periods.

Ovalbumin (OVA) can encapsulate hydrophobic molecules, thus improving aqueous solubility and reducing volatility. For this reason, the potential of an ovalbumin nanocarrier was explored to improve the carvacrol antibacterial properties. The MIC of OVA-Car nanoparticles against *B. cereus* decreased from 0.3875 to 0.0968 mg/mL compared to non-encapsulated carvacrol, while the MIC against *Salmonella* decreased from 0.3875 to 0.1937 mg/mL. These results showed that the encapsulation of carvacrol reduced the concentrations needed to inhibit food pathogens and could be an alternative to maintain the concentration and stability of carvacrol (Rao et al., 2020).

The combination of antibacterial agents with carvacrol is another possibility that food industries could practice. For example, Ni et al. (2020) used bacteriophages (PN05 and PN09) in combination with carvacrol against *Pseudomonas syringae,* a causative agent of damage in kiwifruit. Results demonstrated that carvacrol, in combination with phages, showed better antibiofilm activity than the individual compounds. In another investigation, this monoterpene was combined with sodium dodecyl sulfate, enhancing the antibiofilm activity. Bacteria such as *Lactobacillus fermentum*, *Lactobacillus plantarum*, and *Leuconostoc mesenteroides* cause biofilm formation on the surfaces of ethanol production plants, but the combination mentioned above could remove up to 82% of biofilm biomass (Rigotti et al., 2017). Different antibiofilm agents could be an interesting option for food industries, especially for application on different surfaces.

Carvacrol is an excellent compound for use in the food industry; research has shown that this individual component or combination with others can exert an antibiofilm activity that is relevant. This antibiofilm and antibacterial activity, and its ability to maintain the sensory characteristics of food, make it an excellent option to be used in the food industry.

Conclusion and Future Perspectives

This chapter summarizes the potential use of carvacrol as an antibiofilm agent in the food industry. As evidenced previously, many studies have shown the high antimicrobial activity of carvacrol and its ability to inhibit the biofilm formation of several microorganisms that cause foodborne diseases such as *E. coli*, *S.* Typhimurium, *L. monocytogenes*, and *S. aureus*, among others. Some antibiofilm mechanisms have been proposed for carvacrol, among which the inhibition of QS, motility and EPS synthesis are the main ones. Biofilms are the major cause of cross-contamination problems in the meat, dairy, and fresh produce areas. Therefore, carvacrol is an interesting alternative to the synthetic disinfectants traditionally used in the food industry. However, their application could be limited, considering their high volatility and aroma. Therefore, the studies have focused on encapsulating carvacrol in different matrices such as liposomes, cyclodextrins, nanoparticles, and edible films. There is still much more to investigate to find the best and most efficient ways to apply carvacrol against bacterial biofilms.

References

Abebe, G. M. (2020). The role of bacterial biofilm in antibiotic resistance and food contamination. *International journal of microbiology, 2020*.

Aghaei, Y., Hossein Mirjalili, M., & Nazeri, V. (2013). Chemical diversity among the essential oils of wild populations of *Stachys lavandulifolia* VAHL (Lamiaceae) from Iran. *Chemistry Biodiversity, 10*(2), 262-273.

Ashrafudoulla, M., Mizan, M. F. R., Park, S. H., & Ha, S.-D. (2021). Antibiofilm activity of carvacrol against *Listeria monocytogenes* and *Pseudomonas aeruginosa* biofilm on MBEC™ biofilm device and polypropylene surface. *LWT, 147*, 111575.

Ashrafudoulla, M., Rahaman Mizan, M. F., Park, S. H., & Ha, S.-D. (2021). Antibiofilm activity of carvacrol against Listeria monocytogenes and Pseudomonas aeruginosa biofilm on MBEC™ biofilm device and polypropylene surface. *LWT, 147*, 111575. doi: https://doi.org/10.1016/j.lwt.2021.111575.

Bagci, Y., Kan, Y., Dogu, S., & Çelik, S. A. (2017). The essential oil compositions of *Origanum majorana* L. cultivated in Konya and collected from Mersin-Turkey. *Indian Journal of Pharmaceutical Education and Research, 51*, S463-S469.

Barbosa, L. N., Alves, F. C. B., Andrade, B. F. M. T., Albano, M., Rall, V. L. M., Fernandes, A. A. H., . . . Fernandes Junior, A. (2020). Proteomic analysis and antibacterial resistance mechanisms of Salmonella Enteritidis submitted to the inhibitory effect of Origanum vulgare essential oil, thymol and carvacrol. *Journal of Proteomics, 214*, 103625. doi: https://doi.org/10.1016/j.jprot.2019.103625.

Baydar, H., Sağdiç, O., Özkan, G., & Karadoğan, T. (2004). Antibacterial activity and composition of essential oils from *Origanum*, *Thymbra* and *Satureja* species with commercial importance in Turkey. *Food Control, 15*(3), 169-172.

Brackman, G., & Coenye, T. (2015). Quorum sensing inhibitors as anti-biofilm agents. *Current Pharmaceutical Design, 21*(1), 5-11.

Burt, S. A., Ojo-Fakunle, V. T. A., Woertman, J., & Veldhuizen, E. J. A. (2014). The natural antimicrobial carvacrol inhibits Quorum Sensing in *Chromobacterium violaceum* and reduces bacterial biofilm formation at sublethal concentrations. *PLOS ONE, 9*(4), e93414. doi: 10.1371/journal.pone.0093414.

Cacciatore, F. A., Dalmás, M., Maders, C., Isaía, H. A., Brandelli, A., & da Silva Malheiros, P. (2020). Carvacrol encapsulation into nanostructures: Characterization and antimicrobial activity against foodborne pathogens adhered to stainless steel. *Food Research International, 133*, 109143.

Campana, R., & Baffone, W. (2018). Carvacrol efficacy in reducing microbial biofilms on stainless steel and in limiting re-growth of injured cells. *Food Control, 90*, 10-17.

Can Baser, K. (2008). Biological and pharmacological activities of carvacrol and carvacrol bearing essential oils. *Current Pharmaceutical Design, 14*(29), 3106-3119.

CDC, C. f. D. C. a. P. (2018). *Foodborne outbreaks*. Retrieved August 03, 2021, from https://www.cdc.gov/foodsafety/outbreaks/multistate-outbreaks/outbreaks-list.html

CDC, C. f. D. C. a. P. (2021, July 15, 2021). *Foodborne outbreaks*. Retrieved July 17, 2021, from https://www.cdc.gov/foodsafety/outbreaks/multistate-outbreaks/outbreaks-list.html

Costard, S., Espejo, L., Groenendaal, H., & Zagmutt, F. J. (2017). Outbreak-related disease burden associated with consumption of unpasteurized cow's milk and cheese, United States, 2009–2014. *Emerging infectious diseases, 23*(6), 957.

Côté, H., Pichette, A., St-Gelais, A., & Legault, J. (2021). The biological activity of *Monarda didyma* L. essential oil and its effect as a diet supplement in mice and broiler chicken. *Molecules, 26*(11), 3368.

Coughlan, L. M., Cotter, P. D., Hill, C., & Alvarez-Ordóñez, A. (2016). New Weapons to Fight Old Enemies: Novel strategies for the (bio)control of bacterial biofilms in the food industry. *Frontiers in Microbiology, 7*(1641). doi: 10.3389/fmicb.2016.01641.

Chavan, P. S., & Tupe, S. G. (2014). Antifungal activity and mechanism of action of carvacrol and thymol against vineyard and wine spoilage yeasts. *Food Control, 46*, 115-120. doi: https://doi.org/10.1016/j.foodcont.2014.05.007.

De Vincenzi, M., Stammati, A., De Vincenzi, A., & Silano, M. (2004). Constituents of aromatic plants: carvacrol. *Fitoterapia, 75*(7-8), 801-804.

Delbeke, S., Ceuppens, S., Jacxsens, L., & Uyttendaele, M. (2015). Microbiological analysis of pre packed sweet basil (*Ocimum basilicum*) and coriander (*Coriandrum sativum*) leaves for the presence of *Salmonella* spp. and Shiga toxin-producing *E. coli*. *International Journal of Food Microbiology, 208*, 11-18. doi: https://doi.org/10.1016/j.ijfoodmicro.2015.05.009.

EFSA-ECDC. (2019). The European Union one health 2018 zoonoses report. *EFSA Journal, 17*(12), e05926.

El Gendy, A. N., Leonardi, M., Mugnaini, L., Bertelloni, F., Ebani, V. V., Nardoni, S., . . . Pistelli, L. (2015). Chemical composition and antimicrobial activity of essential oil of wild and cultivated *Origanum syriacum* plants grown in Sinai, Egypt. *Industrial Crops Products, 67*, 201-207.

Espina, L., Berdejo, D., Alfonso, P., García-Gonzalo, D., & Pagán, R. (2017). Potential use of carvacrol and citral to inactivate biofilm cells and eliminate biofouling. *Food Control, 82*, 256-265.

Fachini-Queiroz, F. C., Kummer, R., Estevao-Silva, C. F., Carvalho, M. D. d. B., Cunha, J. M., Grespan, R., . . . Cuman, R. K. N. (2012). Effects of thymol and carvacrol, constituents of *Thymus vulgaris* L. essential oil, on the inflammatory response. *Evidence-Based Complementary Alternative Medicine, 2012*.

Faille, C., Cunault, C., Dubois, T., & Benezech, T. (2018). Hygienic design of food processing lines to mitigate the risk of bacterial food contamination with respect to environmental concerns. *Innovative Food Science & Emerging Technologies, 46*, 65-73.

Friedlander, A., Nir, S., Reches, M., & Shemesh, M. (2019). Preventing biofilm formation by dairy-associated bacteria using peptide-coated surfaces. *Frontiers in microbiology, 10*, 1405.

Friedman, M. (2014). Chemistry and multibeneficial bioactivities of carvacrol (4-Isopropyl-2-methylphenol), a component of essential oils produced by aromatic plants and spices. *Journal of Agricultural and Food Chemistry, 62*(31), 7652-7670. doi: 10.1021/jf5023862.

Fusco, V., Chieffi, D., Fanelli, F., Logrieco, A. F., Cho, G.-S., Kabisch, J., . . . Franz, C. M. A. P. (2020). Microbial quality and safety of milk and milk products in the 21st century. *Comprehensive Reviews in Food Science and Food Safety, 19*(4), 2013-2049. doi: https://doi.org/10.1111/1541-4337.12568.

Galehassadi, M., Rezaii, E., Najavand, S., Mahkam, M., & Mohammadzadeh, G. (2014). Isolation of carvacol from *Origanum vulgare*, synthesis of some organosilicon derivatives, and investigating of its antioxidant, antibacterial activities. *Standard Scientific Research and Essays, 2*, 438-450.

García-Heredia, A., García, S., Merino-Mascorro, J. Á., Feng, P., & Heredia, N. J. F. m. (2016). *Natural plant products inhibits growth and alters the swarming motility, biofilm formation, and expression of virulence genes in enteroaggregative and enterohemorrhagic Escherichia coli. 59*, 124-132.

Ghahramanloo, K. H., Kamalidehghan, B., Javar, H. A., Widodo, R. T., Majidzadeh, K., & Noordin, M. I. (2017). Comparative analysis of essential oil composition of Iranian and *Indian Nigella sativa* L. extracted using supercritical fluid extraction and solvent extraction. *Drug Design, Development Therapy, 11*, 2221.

Ghosh, M., Schepetkin, I. A., Özek, G., Özek, T., Khlebnikov, A. I., Damron, D. S., & Quinn, M. T. (2020). Essential oils from *Monarda fistulosa*: Chemical composition and activation of transient receptor potential A1 (TRPA1) channels. *Molecules, 25*(21), 4873.

Giaouris, E., Heir, E., Hébraud, M., Chorianopoulos, N., Langsrud, S., Møretrø, T., . . . Nychas, G.-J. (2014). Attachment and biofilm formation by foodborne bacteria in meat processing environments: causes, implications, role of bacterial interactions and control by alternative novel methods. *Meat science, 97*(3), 298-309.

Giaouris, E. E., & Simões, M. V. (2018a). Chapter 11 - Pathogenic biofilm formation in the food industry and alternative control strategies. In A. M. Holban & A. M. Grumezescu (Eds.), *Foodborne Diseases* (pp. 309-377): Academic Press.

Giaouris, E. E., & Simões, M. V. (2018b). Pathogenic biofilm formation in the food industry and alternative control strategies *Foodborne Diseases* (pp. 309-377): Elsevier.

Gómez-Sequeda, N., Cáceres, M., Stashenko, E. E., Hidalgo, W., & Ortiz, C. (2020). Antimicrobial and antibiofilm activities of essential oils against *Escherichia coli* O157:H7 and Methicillin-Resistant *Staphylococcus aureus* (MRSA). *Antibiotics, 9*(11), 730.

González-Trujano, M. E., Hernández-Sánchez, L. Y., Muñoz Ocotero, V., Dorazco-González, A., Guevara Fefer, P., & Aguirre-Hernández, E. (2017). Pharmacological evaluation of the anxiolytic-like effects of Lippia graveolens and bioactive compounds. *Pharmaceutical Biology, 55*(1), 1569-1576.

Gutiérrez-Pacheco, M. M., Bernal-Mercado, A. T., Vázquez-Armenta, F. J., Martínez-Tellez, M. A., González-Aguilar, G. A., Lizardi-Mendoza, J., . . . Ayala-Zavala, J. F. (2019). Quorum sensing interruption as a tool to control virulence of plant pathogenic bacteria. *Physiological and Molecular Plant Pathology, 106*, 281-291. doi: https://doi.org/10.1016/j.pmpp.2019.04.002.

Gutierrez-Pacheco, M. M., Gonzalez-Aguilar, G. A., Martinez-Tellez, M. A., Lizardi-Mendoza, J., Madera-Santana, T. J., Bernal-Mercado, A. T., . . . Ayala-Zavala, J. F. (2018). Carvacrol inhibits biofilm formation and production of extracellular polymeric substances of Pectobacterium carotovorum subsp. carotovorum. *Food Control, 89*, 210-218. doi: https://doi.org/10.1016/j.foodcont.2018.02.007.

Han, R., Klu, Y. A. K., & Chen, J. (2017). Attachment and biofilm formation by selected strains of *Salmonella enterica* and entrohemorrhagic *Escherichia coli* of fresh produce origin. *Journal of Food Science, 82*(6), 1461-1466.

Hussain, M. A. (2016). *Food contamination: major challenges of the future*: Multidisciplinary Digital Publishing Institute.

Isoken, H. (2015). Biofilm formation of *Salmonella* species isolated from fresh cabbage and spinach. *Journal of Applied Sciences and Environmental Management, 19*(1), 45-50.

Jahdkaran, E., Hosseini, S. E., Mohammadi Nafchi, A., & Nouri, L. (2021). The effects of methylcellulose coating containing carvacrol or menthol on the physicochemical, mechanical, and antimicrobial activity of polyethylene films. *Food Science & Nutrition, 9*(5), 2768-2778.

Joshi, J. R., Khazanov, N., Senderowitz, H., Burdman, S., Lipsky, A., & Yedidia, I. (2016). Plant phenolic volatiles inhibit quorum sensing in pectobacteria and reduce their virulence by potential binding to ExpI and ExpR proteins. *Scientific Reports, 6*(1), 38126. doi: 10.1038/srep38126.

Kachur, K., & Suntres, Z. (2020). The antibacterial properties of phenolic isomers, carvacrol and thymol. *Critical reviews in food science and nutrition, 60*(18), 3042-3053.

Karaca, B., Buzrul, S., & Coleri Cihan, A. (2019). Anoxybacillus and Geobacillus biofilms in the dairy industry: effects of surface material, incubation temperature and milk type. *Biofouling, 35*(5), 551-560.

Kazemi, M. (2015). Chemical composition and antioxidant properties of the essential oil of *Nigella sativa* L. *Bangladesh Journal of Botany, 44*(1), 111-116.

Khadir, A., Sobeh, M., Gad, H. A., Benbelaid, F., Bendahou, M., Peixoto, H., . . . Wink, M. (2016). Chemical composition and biological activity of the essential oil from *Thymus lanceolatus*. *Zeitschrift für Naturforschung C, 71*(5-6), 155-163.

Krisilia, V., Deli, G., Koutsaviti, A., & Tzakou, O. (2021). *Thymbra* L. and *Satureja* L. essential oils as rich sources of carvacrol, a food additive with health-promoting effects. *American Journal of Essential Oils Natural Products, 9*(1), 12-23.

Kuda, T., Koyanagi, T., Shibata, G., Takahashi, H., & Kimura, B. (2016). Effect of carrot residue on the desiccation and disinfectant resistances of food related pathogens adhered to a stainless steel surfaces. *LWT, 74*, 251-254. doi: https://doi.org/10.1016/j.lwt.2016.07.048.

Kuda, T., Shibata, G., Takahashi, H., & Kimura, B. (2015). Effect of quantity of food residues on resistance to desiccation of food-related pathogens adhered to a stainless steel surface. *Food Microbiology, 46*, 234-238. doi: https://doi.org/10.1016/j.fm.2014.08.014.

Lapointe, C., Deschênes, L., Ells, T. C., Bisaillon, Y., & Savard, T. (2019). Interactions between spoilage bacteria in tri-species biofilms developed under simulated meat processing conditions. *Food micro-biology, 82*, 515-522.

Lee, S. H. I., Barancelli, G. V., de Camargo, T. M., Corassin, C. H., Rosim, R. E., da Cruz, A. G., . . . de Oliveira, C. A. F. (2017). Biofilm-producing ability of Listeria monocytogenes isolates from Brazilian cheese processing plants. *Food Research International, 91*, 88-91.

Lindsay, D., & Von Holy, A. (2006). Bacterial biofilms within the clinical setting: what healthcare professionals should know. *Journal of Hospital Infection, 64*(4), 313-325.

Liu, F., Jin, P., Sun, Z., Du, L., Wang, D., Zhao, T., & Doyle, M. P. (2021). Carvacrol oil inhibits biofilm formation and exopolysaccharide production of *Enterobacter cloacae*. *Food Control, 119*, 107473. doi: https://doi.org/10.1016/j.foodcont.2020.107473.

Liu, N. T., Lefcourt, A. M., Nou, X., Shelton, D. R., Zhang, G., & Lo, Y. M. (2013). Native microflora in fresh-cut produce processing plants and their potentials for biofilm formation. *Journal of food protection, 76*(5), 827-832.

Ma, Z., Bumunang, E. W., Stanford, K., Bie, X., Niu, Y. D., & McAllister, T. A. (2019). Biofilm formation by shiga toxin-producing *Escherichia coli* on stainless steel coupons as affected by temperature and incubation time. *Microorganisms, 7*(4), 95.

Marchese, A., Arciola, C. R., Coppo, E., Barbieri, R., Barreca, D., Chebaibi, S., . . . Daglia, M. (2018). The natural plant compound carvacrol as an antimicrobial and anti-biofilm agent: mechanisms, synergies and bio-inspired anti-infective materials. *Biofouling, 34*(6), 630-656.

Marinelli, L., Di Stefano, A., & Cacciatore, I. (2018). Carvacrol and its derivatives as antibacterial agents. *Phytochemistry Reviews, 17*(4), 903-921.

Mauriello, E., Ferrari, G., & Donsì, F. (2021). Effect of formulation on properties, stability, carvacrol release and antimicrobial activity of carvacrol emulsions. *Colloids and Surfaces B: Biointerfaces, 197*, 111424. doi: https://doi.org/10.1016/j.colsurfb.2020.111424.

Mechergui, K., Coelho, J. A., Serra, M. C., Lamine, S. B., Boukhchina, S., & Khouja, M. L. (2010). Essential oils of *Origanum vulgare* L. subsp. *glandulosum* (Desf.) Ietswaart from Tunisia: chemical composition and antioxidant activity. *Journal of the Science of Food Agriculture, 90*(10), 1745-1749.

Mihajilov-Krstev, T., Radnović, D., Kitić, D., Zlatković, B., Ristić, M., & Branković, S. (2009). Chemical composition and antimicrobial activity of *Satureja hortensis* L. essential oil. *Open Life Sciences, 4*(3), 411-416.

Mittal, R. P., Rana, A., & Jaitak, V. (2019). Essential oils: an impending substitute of synthetic antimicrobial agents to overcome antimicrobial resistance. *Current drug targets, 20*(6), 605-624.

Mohammed, N. K., Manap, A., Yazid, M., Tan, C. P., Muhialdin, B. J., Alhelli, A. M., & Meor Hussin, A. S. (2016). The effects of different extraction methods on antioxidant properties, chemical composition, and thermal behavior of black seed (*Nigella sativa* L.) oil. *Evidence-Based Complementary Alternative Medicine, 2016*.

Møretrø, T., & Langsrud, S. (2004). Listeria monocytogenes: biofilm formation and persistence in food-processing environments. *Biofilms, 1*(2), 107-121.

Morshedloo, M. R., Salami, S. A., Nazeri, V., Maggi, F., & Craker, L. (2018). Essential oil profile of oregano *(Origanum vulgare* L.) populations grown under similar soil and climate conditions. *Industrial Crops Products, 119*, 183-190.

Muhterem-Uyar, M., Dalmasso, M., Bolocan, A. S., Hernandez, M., Kapetanakou, A. E., Kuchta, T., . . . Wagner, M. (2015). Environmental sampling for *Listeria monocytogenes* control in food processing facilities reveals three contamination scenarios. *Food Control, 51*, 94-107. doi: https://doi.org/10.1016/j.foodcont.2014.10.042.

Ni, P., Wang, L., Deng, B., Jiu, S., Ma, C., Zhang, C., . . . Wang, S. (2020). Combined application of bacteriophages and carvacrol in the control of

Pseudomonas syringae pv. actinidiae planktonic and biofilm forms. *Microorganisms, 8*(6), 837.

Nostro, A., & Papalia, T. (2012). Antimicrobial activity of carvacrol: current progress and future prospectives. *Recent patents on anti-infective drug discovery, 7*(1), 28-35.

Novak, J., Lukas, B., & Franz, C. (2010). Temperature influences thymol and carvacrol differentially in *Origanum* spp. (Lamiaceae). *Journal of Essential Oil Research, 22*(5), 412-415.

Orhan-Yanıkan, E., da Silva-Janeiro, S., Ruiz-Rico, M., Jiménez-Belenguer, A. I., Ayhan, K., & Barat, J. M. (2019). Essential oils compounds as antimicrobial and antibiofilm agents against strains present in the meat industry. *Food Control, 101*, 29-38. doi: https://doi.org/10.1016/j.foodcont.2019.02.035.

Popovici, R. A., Vaduva, D., Pinzaru, I., Dehelean, C. A., Farcas, C. G., Coricovac, D., . . . Lazureanu, V. (2019). A comparative study on the biological activity of essential oil and total hydro-alcoholic extract of *Satureja hortensis* L. *Experimental Therapeutic Medicine, 18*(2), 932-942.

Pourhosseini, S. H., Ahadi, H., Aliahmadi, A., & Mirjalili, M. H. (2020). Chemical composition and antibacterial activity of the carvacrol-rich essential oils of *Zataria multiflora* Boiss. (Lamiaceae) from southern natural habitats of Iran. *Journal of Essential Oil Bearing Plants, 23*(4), 779-787.

Raina, A. P., & Negi, K. S. (2012). Essential oil composition of *Origanum majorana* and *Origanum vulgare* ssp. hirtum growing in India. *Chemistry of Natural Compounds, 47*(6), 1015-1017.

Rana, S., Parisi, B., Reineke, K., Stewart, D., Schlesser, J., Tortorello, M., & Fu, T. (2010). *Factors affecting Salmonella cross-contamination during postharvest washing of tomatoes*. Illinois Institute of Technology.

Rao, S.-q., Sun, M.-l., Hu, Y., Zheng, X.-f., Yang, Z.-q., & Jiao, X.-a. (2021). ε-Polylysine-coated liposomes loaded with a β-CD inclusion complex loaded with carvacrol: Preparation, characterization, and antibacterial activities. *LWT, 146*, 111422. doi: https://doi.org/10.1016/j.lwt.2021.111422.

Rao, S., Xu, G., Lu, X., Zhang, R., Gao, L., Wang, Q., . . . Jiao, X. (2020). Characterization of ovalbumin-carvacrol inclusion complexes as delivery systems with antibacterial application. *Food Hydrocolloids, 105*, 105753. doi: https://doi.org/10.1016/j.foodhyd.2020.105753.

Raut, J. S., & Karuppayil, S. M. (2014). A status review on the medicinal properties of essential oils. *Industrial Crops and Products, 62*, 250-264. doi: https://doi.org/10.1016/j.indcrop.2014.05.055.

Requena, R., Vargas, M., & Chiralt, A. (2019). Study of the potential synergistic antibacterial activity of essential oil components using the thiazolyl blue tetrazolium bromide (MTT) assay. *LWT, 101*, 183-190. doi: https://doi.org/10.1016/j.lwt.2018.10.093.

Rigotti, R. T., Corrêa, J. A. F., Maia, N. J. L., Cesaro, G., Rosa, E. A. R., de Macedo, R. E. F., & Luciano, F. B. (2017). Combination of natural antimicrobials and sodium dodecyl sulfate for disruption of biofilms formed by contaminant bacteria isolated from sugarcane mills. *Innovative Food Science & Emerging Technologies, 41*, 26-33.

Rodrigues, J. B. d. S., Souza, N. T. d., Scarano, J. O. A., Sousa, J. M. d., Lira, M. C., Figueiredo, R. C. B. Q. d., . . . Magnani, M. (2018). Efficacy of using oregano essential oil and carvacrol to remove young and mature *Staphylococcus aureus* biofilms on food-contact surfaces of stainless steel. *LWT, 93*, 293-299. doi: https://doi.org/10.1016/j.lwt.2018.03.052.

Rodriguez-Garcia, I., Cruz-Valenzuela, M. R., Silva-Espinoza, B. A., Gonzalez-Aguilar, G. A., Moctezuma, E., Gutierrez-Pacheco, M. M., . . . Ayala-Zavala, J. F. (2016). Oregano (*Lippia graveolens*) essential oil added within pectin edible coatings prevents fungal decay and increases the antioxidant capacity of treated tomatoes. *Journal of the Science of Food Agriculture, 96*(11), 3772-3778.

Rota, M. C., Herrera, A., Martínez, R. M., Sotomayor, J. A., & Jordán, M. J. (2008). Antimicrobial activity and chemical composition of *Thymus vulgaris*, *Thymus zygis* and *Thymus hyemalis* essential oils. *Food Control, 19*(7), 681-687.

Saedi Dezaki, E., Mahmoudvand, H., Sharififar, F., Fallahi, S., Monzote, L., & Ezatkhah, F. (2016). Chemical composition along with anti-leishmanial and cytotoxic activity of *Zataria multiflora*. *Pharmaceutical Biology, 54*(5), 752-758.

Sahraoui, N., Hazzit, M., & Boutekedjiret, C. (2017). Effects of microwave heating on the antioxidant and insecticidal activities of essential oil of *Origanum glandulosum* Desf. obtained by microwave steam distillation. *Journal of Essential Oil Research, 29*(5), 420-429.

Sant'Ana, A. S., Igarashi, M. C., Landgraf, M., Destro, M. T., & Franco, B. D. G. M. (2012). Prevalence, populations and pheno- and genotypic characteristics of *Listeria monocytogenes* isolated from ready-to-eat vegetables marketed in São Paulo, Brazil. *International Journal of Food Microbiology, 155*(1), 1-9. doi: https://doi.org/10.1016/j.ijfoodmicro.2011.12.036.

Sela Saldinger, S., & Manulis-Sasson, S. (2015). What else can we do to mitigate contamination of fresh produce by foodborne pathogens? *Microbial biotechnology, 8*(1), 29-31. doi: 10.1111/1751-7915.12231.

Sharifi-Rad, M., Varoni, E. M., Iriti, M., Martorell, M., Setzer, W. N., del Mar Contreras, M., . . . Tajbakhsh, M. (2018). Carvacrol and human health: A comprehensive review. *Phytotherapy Research, 32*(9), 1675-1687.

Shi, X., & Zhu, X. (2009). Biofilm formation and food safety in food industries. *Trends in Food Science & Technology, 20*(9), 407-413.

Silva, D. A., Botelho, C. V., Martins, B. T., Tavares, R. M., Camargo, A. C., Yamatogi, R. S., . . . Nero, L. A. (2020). Listeria monocytogenes from farm to fork in a Brazilian pork production chain. *Journal of food protection, 83*(3), 485-490.

Sokolik, C. G., Ben-Shabat-Binyamini, R., Gedanken, A., & Lellouche, J.-P. (2018). Proteinaceous microspheres as a delivery system for carvacrol and thymol in antibacterial applications. *Ultrasonics Sonochemistry, 41*, 288-296. doi: https://doi.org/10.1016/j.ultsonch. 2017.09.032.

Soto-Armenta, L., Sacramento-Rivero, J., Acereto-Escoffié, P., Peraza-González, E., Reyes-Sosa, C., & Rocha-Uribe, J. (2017). Extraction yield of essential oil from *Lippia graveolens* leaves by steam distillation at laboratory and pilot scales. *Journal of Essential Oil Bearing Plants, 20*(3), 610-621.

Srey, S., Jahid, I. K., & Ha, S.-D. (2013). Biofilm formation in food industries: a food safety concern. *Food Control, 31*(2), 572-585.

Stratakos, A. C., Sima, F., Ward, P., Linton, M., Kelly, C., Pinkerton, L., . . . Corcionivoschi, N. (2018). The in vitro effect of carvacrol, a food additive, on the pathogenicity of O157 and non-O157 Shiga-toxin producing *Escherichia coli*. *Food Control, 84*, 290-296. doi: https://doi.org/10.1016/ j.foodcont.2017.08.014.

Takahashi, H., Nakamura, A., Fujino, N., Sawaguchi, Y., Sato, M., Kuda, T., & Kimura, B. (2021). Evaluation of the antibacterial activity of allyl isothiocyanate, clove oil, eugenol and carvacrol against spoilage lactic acid bacteria. *LWT, 145*, 111263. doi: https://doi.org/10.1016/j.lwt.2021.111263.

Tapia-Rodriguez, M. R., Bernal-Mercado, A. T., Gutierrez-Pacheco, M. M., Vazquez-Armenta, F. J., Hernandez-Mendoza, A., Gonzalez-Aguilar, G. A., ... Ayala-Zavala, J. F. (2019). Virulence of *Pseudomonas aeruginosa* exposed to carvacrol: alterations of the Quorum sensing at enzymatic and gene levels. *Journal of Cell Communication and Signaling, 13*(4), 531-537. doi: 10.1007/s12079-019-00516-8.

Tapia-Rodriguez, M. R., Hernandez-Mendoza, A., Gonzalez-Aguilar, G. A., Martinez-Tellez, M. A., Martins, C. M., & Ayala-Zavala, J. F. (2017). Carvacrol as potential Quorum Sensing inhibitor of *Pseudomonas aeruginosa* and biofilm production on stainless steel surfaces. *Food Control, 75*, 255-261. doi: https://doi.org/10.1016/j.foodcont.2016.12.014.

Tongnuanchan, P., & Benjakul, S. (2014). Essential oils: extraction, bioactivities, and their uses for food preservation. *Journal of Food Science, 79*(7), R1231-R1249.

Ukuku, D. O., & Fett, W. F. (2006). Effects of cell surface charge and hydrophobicity on attachment of 16 *Salmonella* serovars to cantaloupe rind and decontamination with sanitizers. *Journal of food protection, 69*(8), 1835-1843.

Wagner, E. M., Pracser, N., Thalguter, S., Fischel, K., Rammer, N., Pospíšilová, L., ... Rychli, K. (2020). Identification of biofilm hotspots in a meat processing environment: Detection of spoilage bacteria in multi-species biofilms. *International journal of food microbiology, 328*, 108668.

Walczak, M., Michalska-Sionkowska, M., Olkiewicz, D., Tarnawska, P., & Warżyńska, O. (2021). Potential of carvacrol and thymol in reducing biofilm formation on technical surfaces. *Molecules, 26*(9), 2723.

Wang, Y., Feng, L., Lu, H., Zhu, J., Kumar, V., & Liu, X. (2021). Transcriptomic analysis of the food spoilers *Pseudomonas fluorescens* reveals the antibiofilm of carvacrol by interference with intracellular signaling processes. *Food Control, 127*, 108115. doi: https://doi.org/10.1016/j.foodcont.2021.108115.

Wang, Y., Hong, X., Liu, J., Zhu, J., & Chen, J. (2020). Interactions between fish isolates *Pseudomonas fluorescens* and *Staphylococcus aureus* in dual-species biofilms and sensitivity to carvacrol. *Food Microbiology, 91*, 103506. doi: https://doi.org/10.1016/j.fm.2020.103506.

Warner, J. C., Rothwell, S. D., & Keevil, C. W. (2008). Use of episcopic differential interference contrast microscopy to identify bacterial biofilms on salad leaves and track colonization by *Salmonella* Thompson. *Environmental Microbiology, 10*(4), 918-925. doi: https://doi.org/10.1111/j.1462-2920.2007.01511.x.

Yang, R., Miao, J., Shen, Y., Cai, N., Wan, C., Zou, L., . . . Chen, J. (2021). Antifungal effect of cinnamaldehyde, eugenol and carvacrol nanoemulsion against Penicillium digitatum and application in postharvest preservation of citrus fruit. *LWT, 141*, 110924. doi: https://doi.org/10.1016/j.lwt.2021.110924.

Yaron, S., & Römling, U. (2014). Biofilm formation by enteric pathogens and its role in plant colonization and persistence. *Microbial biotechnology, 7*(6), 496-516.

Yin, B., Zhu, L., Zhang, Y., Dong, P., Mao, Y., Liang, R., . . . Luo, X. (2018). The characterization of biofilm formation and detection of biofilm-related genes in *Salmonella* isolated from beef processing plants. *Foodborne Pathog Dis, 15*(10), 660-667. doi: 10.1089/fpd.2018.2466.

Zgheib, R., El-Beyrouthy, M., Chaillou, S., Ouaini, N., Rutledge, D. N., Stien, D., . . . Iriti, M. (2019). Chemical variability of the essential oil of *Origanum ehrenbergii* boiss. From Lebanon, assessed by independent component analysis (ICA) and common component and specific weight analysis (CCSWA). *International Journal of Molecular Sciences, 20*(5), 1026.

Zhao, Y., Yang, Y.-H., Ye, M., Wang, K.-B., Fan, L.-M., & Su, F.-W. (2021). Chemical composition and antifungal activity of essential oil from Origanum vulgare against Botrytis cinerea. *Food Chemistry, 365*, 130506. doi: https://doi.org/10.1016/j.foodchem.2021.130506.

Zheng, S., Bawazir, M., Dhall, A., Kim, H.-E., He, L., Heo, J., & Hwang, G. (2021). Implication of surface properties, bacterial motility, and hydrodynamic conditions on bacterial surface sensing and their initial adhesion. *Frontiers in Bioengineering and Biotechnology, 9*, 82.

Zou, M., & Liu, D. (2018). A systematic characterization of the distribution, biofilm-forming potential and the resistance of the biofilms to the CIP processes of the bacteria in a milk powder processing factory. *Food Research International, 113*, 316-326.

Zwirzitz, B., Wetzels, S. U., Dixon, E. D., Stessl, B., Zaiser, A., Rabanser, I., . . . Strachan, C. (2020). The sources and transmission routes of microbial populations throughout a meat processing facility. *NPJ biofilms and microbiomes, 6*(1), 1-12.

Chapter 4

Carvacrol Emulsification: From Theory to Applications

Alana G. de Souza and Derval S. Rosa[*]

Center for Engineering, Modeling, and Applied Social Sciences (CECS),
Federal University of ABC (UFABC)
Santo André, Brazil

Abstract

Carvacrol [2-methyl-5-(1-methylethyl)-phenol] is a monoterpene found in essential oils of aromatic plants with numerous active properties: antibacterial, antioxidant, antifungal, and anti-cancer. However, essential oils have high volatility, hydrophobicity, intense aroma, photosensitivity, low stability, high susceptibility to oxidation, and poor water solubility, limiting their application. Emulsification is considered a simple way to overcome these limitations. Emulsions are formed by homogenization of two immiscible liquid phases in which one of them is dispersed in the other in the form of tiny droplets. Emulsion systems are usually applied in hydrophilic-lipophilic mixtures, such as essential oil and water, to improve the oil solubility and facilitate application. However, an emulsification agent is needed to break the drops of the oily phase into smaller and more stable particles since oil and water do not coexist harmoniously in the same system. Oil molecules have instantaneous dipoles related to the molecule's chemical structure and functional groups, which reorient themselves all the time, generating constant forces of attraction or dispersion. One way to generate emulsions and overcome the energy barrier of the system

---

[*] Corresponding Author's E-mail: dervalrosa@yahoo.com.br.

In: A Closer Look at Carvacrol
Editor: Zak A. Cunningham
ISBN: 978-1-68507-627-6
© 2022 Nova Science Publishers, Inc.

associated with interfacial tension is through shear energy, seeking to balance the system's free energy. However, the stability is low, and there is a reversibility trend, resulting in coalescence in short periods. Additives that reduce the interfacial tension and the energy barrier, such as surfactants and solid particles, are an alternative to overcome this energy barrier and make the emulsions thermodynamically favored. The additives are adsorbed to oil/water interfaces, forming thick interfacial layers to stabilize the emulsions via electrostatic or steric mechanisms, resulting in droplets with micro or nanosizes with high stability. This chapter describes the theoretical aspects of micro and nanoemulsions formation, considering fluid properties, physical and chemical characteristics, and thermodynamic equilibrium. In addition, we do an in-depth review of the emulsification of essential oils that contain carvacrol, addressing various preparation methodologies and types of stabilizers. The solid particles, which make up the well-known Pickering emulsions, will be highlighted among the stabilizers. The state of the art of new technologies and applications is widely presented and discussed, such as for food, biomedical, delivery carrier systems, among others.

Keywords: essential oil, emulsion, nanoparticles

Introduction

Carvacrol [2-Methyl-5-(1-methylethyl) phenol] is a monoterpenoid phenol found in several plants, such as pepperwort, thyme, oregano, marjoram, and bergamot, approved by the Council of Europe and the Food and Drug Administration (Code of Federal Regulations 21 Part 172) as safe for humans (Shrestha et al. 2019). This bioactive compound is a common component of essential oils and has been widely explored for industrial, pharmaceutical, biomedical, and animal industries as antibiotic alternatives due to its excellent antimicrobial activity. However, several challenges are reported concerning carvacrol application due to its water insolubility, high volatility, pungent odor, and high tendency to oxidate or degrade in the environment, reducing its stability and bioactivity (P. Wang and Wu 2021). Several approaches have been studied to improve the application of essential oils, such as encapsulation and emulsification (Mauriello, Ferrari, and Donsì 2021).

Encapsulation is a common technique to alleviate the instability drawbacks and enhance phytochemicals bioavailability and half-life through active compounds protect against environmental factors, such as moisture, light, oxygen, and pH (Hussein, El-Banna, et al. 2017). Recent efforts have led to emulsion systems to improve essential oils' stability and solubility. Emulsions are oil and water mixtures formed using high-intensity energy methods (ultrasound, microfluidization, or high-pressure homogenization) or low energy techniques, such as solvent diffusion. An emulsion generally requires a stabilizer agent to avoid instability and droplets coalescence, commonly a surfactant, amphiphilic molecules that provides an elastic barrier against droplets collision. Besides, based on the droplet size, emulsions are classified in macroemulsion (1-100 μm), microemulsion (10 nm – 100 nm), and nanoemulsion (20 nm – 500 nm) (McClements 2012). Recently, a new classification was introduced in the emulsion's universe, called Pickering emulsion, stabilized by solid particles (McClements and Gumus 2016).

The emulsions are kinetically stable systems that protect the active constituent, i.e., carvacrol, increasing bioactivity, antimicrobial properties, antioxidants, and pharmaceutical uses. This chapter describes the different emulsification systems used to stabilize essential oils containing carvacrol as active compounds, considering the main techniques used to form the emulsions, properties, and possible applications using stabilized emulsions-based products.

Literature Review

Emulsions Classifications

An emulsion can be defined as a two-phase system consisting of two immiscible liquids, one of which (the dispersed phase) is finely and evenly dispersed as droplets throughout the second phase (the continuous phase). Emulsions are widely used in our daily life, such as milk, lotions, and creams, being of great interest in areas of study such as chemistry, biology, physics, and engineering (Zou, Sipponen, and Österberg 2019; Abdul-razzaq, Jaafar, and Bandyopadhyay 2020; Haque et al. 2017; Prado, Gonzales, and Spinacé 2019). In addition, they are of great interest to the pharmaceutical, cosmetology, food, water-based paint industries, among others. In emulsified form, active agents can be diluted to an appropriate level and be more easily applied. However, strict control of the emulsion's process, characteristics, and final properties is necessary to ensure proper applicability.

When an aqueous phase contains dispersed oil droplets, it is called an oil-in-water (O/W) emulsion and, conversely, when the emulsion system is based on an oil phase with dispersed water droplets, it is called a water-in-oil (W/O) (Keivani Nahr et al. 2019; Zhang et al. 2017; Rosen and Kunjappu 2012;

Mendes et al. 2020; Scharamm 2015). Emulsified systems have four main characteristics: i) improved stability for particle aggregation and eventual sedimentation; ii) by virtue of the appropriate small size of the droplets, they only weakly scatter the light wave and are therefore advantageous for adding value to products that need to be optically transparent (or only slightly cloudy); iii) emulsions can be tailored to have new rheological properties; iv) have greater bioavailability of specific types of bioactive molecules contained in the dispersed phase (Bhattacharjee 2019).

In 1993, Walstra described an emulsion as a mixture of oil, water, surfactant, and energy, and the latter needed to break down oil phase droplets into smaller, more stable particles (Walstra 1993). Particle size and distribution are essential characteristics that affect emulsion stability, directly related to rheology and stability for storage or under shear (Goodarzi and Zendehboudi 2019). Particle size is even a classification factor for emulsions, divided into a) macroemulsions, b) microemulsions and c) nanoemulsions.

Macroemulsions are conventional and easy-to-produce emulsions with droplet sizes significantly higher than micro or nanoemulsions. These emulsions are thermodynamically unstable and coalesce and phase out due to reduced interfacial energy over time. Thus, studies focus on the susceptibility to coalescence through the use of stabilizers. Large amounts of surfactant are used to form an elastic mechanical barrier for drop-drop collision (P. Wang and Wu 2021). However, these emulsions tend to have high toxicity due to their high surfactant content and energy expenditure.

Furthermore, due to the higher drop sizes and high interfacial energy, macroemulsions are generally not thermodynamically stable and are sensitive to changing environmental parameters such as pH and ionic strength (Goodarzi and Zendehboudi 2019; Posocco et al. 2016). Syed et al. (2020) prepared carvacrol emulsions with a droplet size of 200 nm and evaluated the antibacterial activity of this compound on raw goat meat surface during extended storage. The authors reported high antimicrobial activity against *Bacillus cereus* and *Escherichia coli* due to the prolonged retention of an adequate concentration of carvacrol compound, resulting in a controlled release throughout the incubation period (Syed, Banerjee, and Sarkar 2020).

Microemulsions are transparent, thermodynamically stable colloidal dispersions, with droplet sizes ranging between 10 and 100 nm. They are simple to manufacture and spontaneously formed through molecular associations, with no input of energy. It uses an intermediate concentration of surfactant, usually ranging from 15 to 30 wt% of the oil phase, which is a limitation due to toxicity and flavor issues.

Furthermore, they are sensitive to changes in temperature and salinity (McClements 2012). Dantas et al. (2021) developed carvacrol microemulsions with droplet diameters of ~130 nm aiming for immunomodulatory action. The emulsions did not show cytotoxicity and improved the biological action by inhibiting the proinflammatory cytotoxins (Dantas et al. 2021).

Nanoemulsions are also considered thermodynamically unstable with droplets with less than 100 nm of diameters. These emulsions are attractive due to their high stability over time, which could guarantee a storage time of months or years and require low surfactant contents (1-3%), i.e., low toxicity (McClements 2020). However, this system generally shows droplets aggregation due to the Ostwald ripening. Hussein et al. (2017) prepared carvacrol-based nanoemulsions stabilized with 3 wt% of Tween 80, a non-ionic surfactant. The authors reported droplet diameters of 39 nm and high monodispersed sizes (Hussein, El-Bana, et al. 2017). The developed emulsions showed potential in attenuating hyperglycemia and neurodegenerative diseases.

In addition to the droplet size classification, a new emulsions class emerged in the last decade that differs from the others due to the emulsifying agent. Pickering emulsions (PEs) are emulsions stabilized by solid particles and, due to the absence of surfactants, have low toxicity, being attractive for a broad range of applications. Among the numerous particles that can be used, polysaccharides are interesting due to their low cost, availability, and sustainability, such as cellulose, chitosan, starch, pectin, and others (McClements and Gumus 2016; Dong et al. 2021; X. Y. Wang and Heuzey 2016; Sufi-Maragheh et al. 2019; Jiang et al. 2019). PEs show high deformation resistance due to the irreversible adsorption of solid particles at the O/W or W/O interface (Low et al. 2020). Zhou et al. (2018) prepared oregano essential oil Pickering emulsions stabilized with cellulose nanocrystals and reported good stability and antimicrobial efficacy against four food-related bacteria due to the presence of carvacrol active compound (Zhou et al. 2018).

Emulsions Formation

According to Walstra, three processes always occur during emulsification: 1) oil droplets deform and break 2) emulsifiers are transported and absorbed by oil droplets along with the formation of oil droplets coated with an emulsifier, and 3) emulsion droplets collide and merge (Walstra 1993). High or low energy methods can carry out the emulsification process.

High energy methods comprise high-pressure homogenization, microfluidization, and ultrasonication and are the most common for emulsions containing essential oils. Low energy methods involve transition phase inversion (phase inversion temperature and phase inversion composition), catastrophic phase inversion (emulsion inversion point), and autonanoemulsification (Kumar et al. 2019). Besides, in addition to the energy, emulsifiers are added to the emulsion to improve its stability since they facilitate the formation of individual oil droplets during homogenization due to disruptive forces (cavitation, turbulence, and shear), which break the larger droplets into smaller droplets, and increase stability, preventing droplets rejoining and forming two distinct phases of oil and water (McClements and Gumus 2016).

High-energy methods utilize mechanical energy, which provides disruptive forces capable of breaking large droplets into smaller droplets, resulting in emulsions with high kinetic energy. The high-pressure methods are interesting due to their flexibility of use since they can be used with highly viscous oils and high molar mass. In addition, this method allows the use of lower surfactant concentrations (Safaya and Rotliwala 2020). Several methods are used to prepare carvacrol emulsions, as described in Table 1.

During emulsification, oil molecules, which have instantaneous dipoles related to the molecule's chemical structure and functional groups, reorient themselves all the time, generating constant forces of attraction or dispersion. The attraction results in the phenomenon of spontaneously minimizing the surface area. According to the emulsification theory, all liquids tend to take a shape that produces less surface area (larger droplet size or complete phase separation). There are three prevalent emulsification theories: oriented wedge, interfacial film, and surface tension.

Table 1. Selected studies of carvacrol emulsions and their preparation methods

Oil/compound	Emulsion composition	Preparation method and conditions	Main results	Reference
Geraniol and carvacrol	Geraniol, carvacrol, tween 80, and medium-chain triglyceride	Ultrasonic emulsification, 20 kHz, 40% amplitude, 5 min	antimicrobial efficacy of geraniol and carvacrol for nine days and high stability	(Syed, Banerjee, and Sarkar 2020)
Carvacrol	Carvacrol, tween 20	High shear homogenization (24000 rpm for 10 min) followed by high-pressure homogenization, ten times, 350 MPa	carvacrol nanoemulsion ameliorate negative changes in the thioacetamide injected rats	(Hussein, El-Banna, et al. 2017)
Carvacrol	Carvacrol, tween 80, chitosan	Mechanical homogenizer (21500 rpm for 5 min) followed by high-pressure homogenization, 100 MPa	Homogeneous emulsion microstructure obtained by combining both preparation techniques	(Flores et al. 2021)
Carvacrol	Carvacrol, n-hexane, Tween 80	Sonicator, 20 kHz, one h	Small diameters (40 nm), high monodispersity, and emulsions increase in fasting blood sugar with a reduction of insulin in diabetic rats	(Hussein, El-Banna, et al. 2017)
Carvacrol	Carvacrol, Tween 80, Span 80	Ultrasonic emulsification, 300 W, 1 min	Good stability over 90 days and bactericidal effect against Escherichia coli, P. aeruginosa, and Salmonella spp.	(Motta Felicio et al. 2021)

Oil/compound	Emulsion composition	Preparation method and conditions	Main results	Reference
Carvacrol and corn oil	Carvacrol, corn oil, and tween 80	High shear mixer, 8000 rpm, 10 min, followed by high-pressure homogenization at 1000 or 100 bar or ultrasound at 20 kHz and 100% amplitude	Stability over one month. Ultrasonication produced nanoemulsions with 309 nm and antimicrobial activity against E. coli and Pichia pastoris.	(Sow et al. 2017)
Mexican oregano oil	Mexican oregano oil, medium-chain triglyceride, tween 80, gum Arabic, hydroxylated soy lecithin Emulfluid HL 66	High shear disperser (9500 rpm for 5 min) followed by ultrasonic processor, 50% amplitude, 15 min	The oil showed 14.6% of carvacrol, and the droplet sizes followed the next descendent order, depending on the stabilizer: gum arabic > lecithin > Tween 80.	(Herrera-Rodríguez et al. 2019)
Carvacrol	Carvacrol, non-ionic surfactant, emulsifier polysorbate 80	Sonication, 25 kHz, 750 W, 10 min	Emulsions showed antitumor potential in vivo using an athymic nude mice model.	(Khan et al. 2018)

The oriented wedge theory assumes that some emulgents orient themselves on the surface and inside the liquid, preferentially soluble in one of the phases, resulting in greater penetration depth and toughness. Depending on the shape and size of the emulgent molecules and their solubility characteristics, a preferential orientation occurs, resulting in a wedge-shaped arrangement around the formed droplets. The interfacial film theory describes that the emulgent is found at the interface between oil and water as a thin layer of a film adsorbed on the surface, preventing droplets' approach and junction. On the other hand, the third theory considers that joining two immiscible liquids generates a resistant force that prevents the liquid from fragmenting into smaller droplets, called interfacial tension (Anton, Benoit, and Saulnier 2008).

The interfacial tension (γ) is proportional to the work of the system (W), from a thermodynamic point of view, according to Equation 1, where ΔA is the interfacial area.

$$W = \gamma \, \Delta A \tag{1}$$

Thus, the system energy is related to the emulsion's average droplet radius. The use of surfactants or solid particles influences the interfacial tension, facilitating the formation of smaller droplets. Mauriello et al. (2021) investigated the formulation effect on emulsions properties, stability, and carvacrol release. The authors investigated Tween 80, whey protein isolates, and Pickering whey protein microgel particles as emulsifiers, and the emulsions were prepared using a high-intensity energy method, using a rotor-stator homogenizer (Mauriello, Ferrari, and Donsì 2021). The lowest interfacial tension was verified for the Pickering emulsions and was ascribed to the molecular interactions between the carvacrol and the solid particles, and this system also showed higher carvacrol solubilization and antimicrobial activity. This study contributed to design emulsions properties based on formulation parameters, demonstrating the relevance of selecting the emulsifier type.

For efficient emulsification, the surfactant or solid particle must reduce surface tension, and this characteristic depends on molecular weight, number and location of hydrophilic groups, wettability with both liquid phases, characteristic of functional groups, among others (McClements and Jafari 2018). Besides, the emulsifier adsorption rate must be faster than the drop fragmentation rate. Otherwise, the droplets will not be fully coated with the emulsifier before a drop breakage.

The tendency of an emulsifier to adsorb at an oil-in-water interface is described using the Gibbs adsorption isotherm, which relates the emulsifier concentration at the interface (Γ) to the interfacial tension and the quantity of emulsifier in the bulk solution (c), according to Equation 2.

$$\Gamma = -\frac{1}{pRT}\left(\frac{d\gamma}{d\ln(c)}\right) \qquad (2)$$

Also, R is the gas constant, p is a parameter that depends on the emulsifier ionic properties, and T is the absolute temperature. For neutral molecules, p = 1, and for charged molecules, p > 1 depending on the counter-ions number released when the emulsifier adsorbs to the interface. Typically, the maximum values of Γ are found when the emulsifier achieves saturation. Using the plot of γ versus ln(c), Γ is the slope of the curve; besides, using the same plot is also possible to obtain the surface activity measure (S_A), where $S_A = 1/C_{1/2}$.

The Langmuir adsorption isotherm is used to describe the fraction of a water-oil interface that is covered by an emulsifier, following Equation 3, where θ is the interface adsorption sites fraction occupied by emulsifier molecules, Γ_∞ is the surface excess concentration when the emulsifier is in excess, and $c_{1/2}$ is the emulsifier concentration in the aqueous phase where $\theta = ½$.

$$\theta = \frac{\Gamma}{\Gamma_\infty} = \frac{c/c_{1/2}}{1+c/c_{1/2}} \qquad (3)$$

The Gibbs and Langmuir adsorption isotherms were the first equations developed to describe emulsions, and at this moment, several authors reported other equations that consider other factors, such as type of emulsifier, orientation, conformational changes, interactions, morphology, and others (Fainerman, Lucassen-Reynders, and Miller 1998; Fainerman et al. 2016; Kotsmar et al. 2009; Bhattacharjee 2019).

The emulsion thermodynamic stability is governed by the change in the free energy of a system. In nanoemulsions, the system is considered thermodynamically unstable, i.e., the emulsion free energy is greater than the free energy of the separate phases (oil and water). The free energy change associated with the dispersion formation consists of an interfacial free energy term (ΔG_{eu}) and a configuration entropy term ($-T\Delta S_{config}$), following Equation 4.

$$\Delta G_{formation} = \Delta G_{eu} - T\Delta S_{config} \qquad (4)$$

The configuration entropy depends on the number of arrangements accessible to the oil phase in the emulsified state and is much higher than in the non-emulsified state, favoring the dispersion formation (Bhattacharjee 2019).

Two main factors determine the kinetic stability: i) Energy barriers: any energy barrier (or activation energy) that separates the two states (final and initial) will determine the rate of conversion from a state to other. The energy barrier height depends on forces operating in the interface of two drops, involving repulsive forces and hydrodynamic interactions (steric and electrostatic); ii) Mass transport phenomena: droplets in an emulsion are particularly unstable concerning growth over time by a process known as Ostwald ripening, in which solute (or mass) molecules are exchanged between droplets by through molecular diffusion through the solvent.

The physicochemical properties of the interfacial layer formed when the emulsifier molecules are adsorbed on an oil-water interface largely determine the stability and functionality of the emulsions, which can vary in thickness, density and rheology depending on molecular dimensions, packaging, and interactions of the adsorbed emulsifier. The interface thickness strongly impacts the strength and amplitude of steric interactions between the emulsion droplets, while the electrical characteristics have a strong influence on the strength of the electrostatic interactions (Zhu et al. 2019; Kralova and Sjöblom 2009).

Another critical point in developing stable emulsions is that the Hydrophilic-Lipophilic Balance (HLB) value plays a vital role in proper emulsification when using surfactants as a stabilizer. In solid particles, the property that governs the stabilization mechanisms is the wettability of the solid particle. The HLB number describes the surfactant solubility and partition characteristics (Bhattacharjee 2019). Surfactants with low HLB (2-6) break down in the oil phase and stabilize water-in-oil emulsions, while those with high HLB (8–18) break down in the water phase and stabilize oil-in-water emulsions, as illustrated in Figure 1.

The contact angle provides wetting magnitude for solid particles as emulsifiers (Xiao, Li, and Huang 2016). According to Kaptay (2006), O/W emulsions are produced when the contact angle is $15°<\theta<90°$ and W/O emulsions are prepared if the contact angle is $90°<\theta<150°$, considering a monolayer of solid particles (Kaptay 2006). The irreversible adsorption of solid particles occurs if the contact angle is $30°<\theta<150°$ since the particles' desorption energy is higher than the Brownian motion energy (Xiao, Li, and Huang 2016).

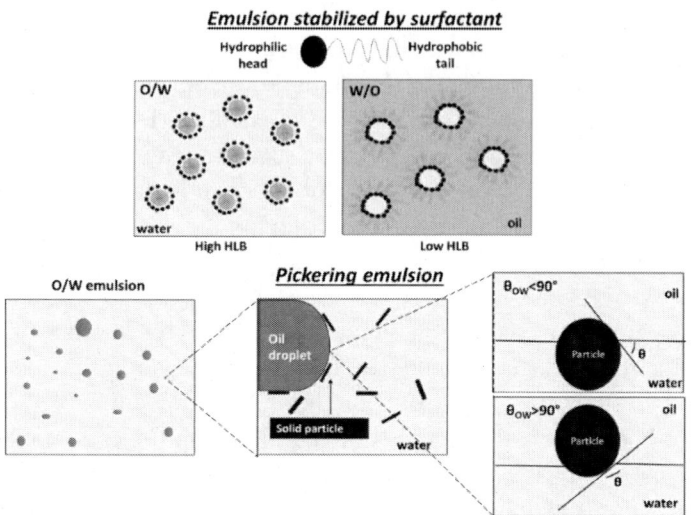

Figure 1. Illustration of emulsions stabilized with surfactants and solid particles and their relations with HLB balance and contact angle, respectively.

Emulsions Properties

The droplet size is a significant characteristic governing the emulsion quality and stability, related to homogeneity and size distribution. The literature presents several techniques to measure the droplet sizes according to the size range, such as laser diffraction, microscopic visualization, and mathematical estimation. Moraes-Lovison et al. (2017) measured the droplet size of oregano nanoemulsions using quasi-elastic light scattering equipment, and the measured diameters ranged between 35 and 65 nm (Moraes-Lovison et al. 2017). Leclercq et al. (2020) developed cyclodextrin-based Pickering emulsions containing carvacrol essential oil. The emulsions showed average droplet sizes between 35 and 39 μm, measured using a light microscope (Leclercq et al. 2020).

The droplet size is associated with interfacial and colloidal properties and emulsion stability, and these properties are generally investigated by monitoring the interfaces and morphology of the droplets. Several techniques are used to investigate these characteristics, mainly microscopic methods that include optical microscopy, phase contrast microscopy, confocal laser scanning microscopy, Raman microscopy, scanning electron microscopy, transmission electron microscopy, and atomic force microscopy (McClements 2007). Figure 2a shows an optical micrography obtained for Pickering emulsions stabilized by cellulose nanofibers with droplet diameters between 50 and 800 μm, indicating emulsion instability due to the high heterogeneity and tendency to coalescence.

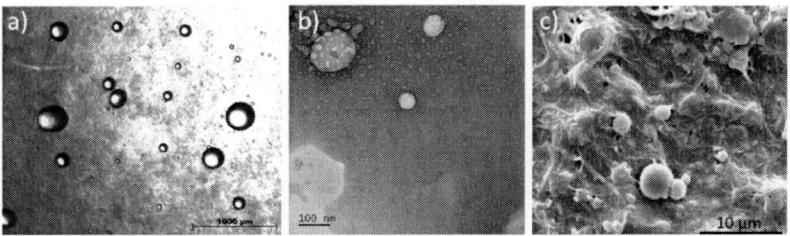

Figure 2. a) Pickering emulsion optical microscopy image, b) nanoemulsion transmission electron photomicroscopy, and c) dried cellulose nanofibers-stabilized Pickering emulsion scanning electron photomicroscopy.

Table 2. Selected studies of carvacrol emulsions and their properties

Emulsion	Particle size	Zeta potential (mV)	Reference
Carvacrol stabilized with tween 80	220 nm	-24	(Syed, Banerjee, and Sarkar 2020)
Carvacrol:sunflower (50:50) stabilized with whey protein isolates	426.7 nm	-12.7	(Mauriello, Ferrari, and Donsì 2021)
Carvacrol nanoemulsion stabilized with tween80/span80	165.7 nm	-10.2	(Dantas et al. 2021)
Carvacrol nanoemulsion stabilized with span 80 or tween 80	~170 nm	Between -14.6 and -39.0	(Motta Felício et al. 2021)
Carvacrol stabilized with tween 80	164.5 nm	-	(Yang et al. 2021)

Figure 2b shows a transmission electron image of nanoemulsions with 100 nm droplet size. Figure 3c shows a scanning electron microscopy image of dried Pickering emulsion stabilized with cellulose nanofibers, where it is possible to identify well-defined capsules, indicating an essential oil encapsulation.

The emulsion stability is another essential parameter for evaluating emulsions quality by referring to their capacity to resist physicochemical changes during the time (McClements 2007). The stability is dependent on droplets morphology, size, rheology, and other physical characteristics along the time. Several instability mechanisms can be observed in emulsions, such as creaming, sedimentation, coalescence, flocculation, Ostwald ripening, and phase separation. Commonly, the stability is evaluated by storage analysis during at least a month, micromanipulation, accelerated coalescence, light diffusion, and centrifugation (Low et al. 2020). Other techniques are also used to characterize emulsions properties, such as nuclear magnetic resonance (NMR), which provides information about droplets size and shape and can be used in complex mixtures, rheology, near-infrared spectroscopy (NIR), a simple technique used to obtain physical and chemical characteristics of samples, Zeta potential, widely used to obtain emulsions' charges, among others (Goodarzi and Zendehboudi 2019). Table 2 shows literature studies of carvacrol emulsions and some relevant properties.

Emulsions Applications

The carvacrol emulsions advanced properties have increased the scientific and industrial interest in developing new products and technologies based on emulsions systems. The main sectors involved in these continuous advances are pharmaceutics and cosmetics, followed by food, pesticide, and medicine (Safaya and Rotliwala 2020). Among the main applications, the food and medicine sectors are highlighted. Carvacrol is widely used in biomedical applications due to its anti-cancer, anti-inflammatory, antioxidant, antidepressant, anti-helminthic, antinociceptive, and antimicrobial activities.

Hussein et al. (2017) investigated carvacrol nanoemulsions aiming for medical applications. The authors reported that carvacrol emulsions could ameliorate the detection of liver enzymes and hepatotoxicity, indicating that carvacrol has potential against hepatic damage induced by thioacetamide. The emulsification improved the bioavailability of carvacrol, suggesting that this form can be used as a protecting agent against liver fibrosis (Hussein, El-Banna, et al. 2017). Hussein et al. (2017) prepared carvacrol nanoemulsions for treating neurodegenerative disorders in experimental diabetes. Diabetes reduces the activity of the antioxidant enzymes and improves the production of reactive oxygen species, inducing oxidative stress in the brain. The use of carvacrol nanoemulsions improved the drug delivery and partially reversed neuronal injury by reducing oxidative stress and improving antioxidant parameters (Hussein, El-Bana, et al. 2017). Herrera-Rodríguez et al. (2019) developed Mexican oregano emulsions with antifungal activity against *Candida albicans*, a pathogen that causes systemic infections in immunocompromised hosts and has significant mortality worldwide. The authors reported that carvacrol's biological properties disturb and damage microorganisms' cytoplasmic and protein membranes, resulting in their death. The results indicate the excellent capacity of nanoemulsions to diffuse and disrupt the *C. Albicans* membrane, resulting in excellent antifungal activity (Herrera-Rodríguez et al. 2019). Dantas et al. (2021) prepared carvacrol nanoemulsions with immunomodulatory activity by evaluating levels of immunosuppressive cytokines, inflammatory mediators, and proinflammatory cytokines. The authors reported that treatment with the developed nanoemulsions for two months reduced the inflammatory cytokine levels and increased the anti-inflammatory cytokines. The droplet size was associated with the emulsion effectiveness since lower diameters can increase the contact surface and biological interaction (Dantas et al. 2021).

Another common application of carvacrol emulsions is in the food industry. Food contamination with microbial pathogens is an urgent concern due to the foodborne diseases mainly caused by bacteria and their toxins, viruses, and other parasites. In many countries over the past two decades, these diseases have been seen as a growing economic and public health problem, with outbreaks of cases attracting media attention and increasing consumer interest. The problem is predicted to increase in the 21st century, especially as various global changes influence international food security, including population growth, poverty, food, and animal feed exports. Several approaches have been studied to improve food quality and safety, and carvacrol emulsions are recognized as efficient growth inhibitors of food pathogens (Motta Felício et al. 2021).

Yang et al. (2021) developed carvacrol nanoemulsions with antifungal effects for citrus fruit preservation. The emulsions inhibited the spore's germination and increased the cell membrane permeability, resulting in fungal death. Besides, nanoemulsions coating reduced the fruits' weight loss and respiratory rate during storage, and antioxidant enzyme activity, measured by SOD, CAT, POD, and APX activities, significantly increased, indicating inhibition of diseases infection on citrus fruits (Yang et al. 2021). Sow et al. (2017) investigated carvacrol nanoemulsions to inactivate *E. coli* and *Pichia pastoris* on shredded cabbages. The authors reported that carvacrol nanoemulsions could be used to sanitize stainless steel surface was food is manipulated, improving fresh food safety (Sow et al. 2017). Motta Felício et al. (2020) developed carvacrol emulsions as antimicrobial bactericidal natural compounds in fresh or processed food. The authors tested the nanoemulsion against *E. coli, P. aeruginosa,* S. Enteritidis, S. Typhimurium, and *S. aureus* (Motta Felício et al. 2021).

Conclusion

Carvacrol is a monoterpene found in various aromatic plants and biosynthesized from γ-terpinene and ρ-cymene. Its high volatility and low water solubility are the main disadvantages limiting essential oil use in commercial products since it is easily oxidized and loses efficiency. Emulsification is a common approach used to overcome these limitations and improve the carvacrol applicability. Emulsions can be classified into macroemulsions, microemulsions, or nanoemulsions, based on the size of the droplets, or even into Pickering emulsions, when the stabilizer used is a solid particle. Emulsification methods are diverse and cover low or high energy techniques, with high stability essential oil emulsions generally requiring high energy to ensure the breakdown of the oil phase into smaller droplets. During emulsification, several physicochemical processes occur to guide the surfactant or solid phase at the interface between liquids, aiming to achieve a thermodynamic balance that guarantees the physical stability of the emulsions. Several researchers have described these processes throughout history, as in the adsorption theory of Gibbs or Langmuir or the free energy equation of the entropy-dependent system. Emulsification is directly related to the properties of emulsions, the main ones being droplet size, morphology, and stability. These characteristics are impacted by other properties such as surface charges, chemical interactions between molecules, rheology, and also govern controlled release mechanisms and are associated with possible applications of carvacrol emulsions. Due to the bioactive properties of carvacrol, several applications are reported, emphasizing medical and food products due to the anti-cancer, anti-inflammatory, antioxidant, antidepressant, anti-helminthic, antinociceptive, and antimicrobial activities. There are also several possibilities for developments in carvacrol emulsions due to their exceptional properties, which can be combined with other essential oils or stabilized with sustainable solid particles to reduce environmental impacts. Furthermore, mechanisms and in-depth understandings of release and stability must be systematically studied, mainly in nanoemulsions and Pickering emulsions, aiming to develop advanced systems for diverse applications.

References

Abdul-razzaq, Rafat, Mohd Zaidi Jaafar, and Sulalit Bandyopadhyay. 2020. "Investigating Synergistic Effects of Surfactants and Nanoparticles (NPs) on

Emulsion Viscosity." *Asian Journal of Fundamental and Applied Sciences* 1 (1): 9–16.
Anton, Nicolas, Jean Pierre Benoit, and Patrick Saulnier. 2008. "Design and Production of Nanoparticles Formulated from Nano-Emulsion Templates-A Review." *Journal of Controlled Release* 128 (3): 185–99. doi:10.1016/j.jconrel.2008.02.007.
Bhattacharjee, Kaustav. 2019. "Importance of Surface Energy in Nanoemulsion." In *Nanoemulsions - Properties, Fabrications and Applications*, edited by Kai Seng Koh, 1:1–20. IntechOpen. doi:10.1016/j.colsurfa.2011.12.014.
Dantas, Amanda Gabrielle Barros, Rafael Limongi de Souza, Anderson Rodrigues de Almeida, Francisco Humberto Xavier Júnior, Maira Galdino da Rocha Pitta, Moacyr Jesus Barreto de Melo Rêgo, and Elquio Eleamen Oliveira. 2021. "Development, Characterization, and Immunomodulatory Evaluation of Carvacrol-Loaded Nanoemulsion." *Molecules* 26 (13): 3899. doi:10.3390/molecules26133899.
Dong, Hui, Qijun Ding, Yifei Jiang, Xia Li, and Wenjia Han. 2021. "Pickering Emulsions Stabilized by Spherical Cellulose Nanocrystals." *Carbohydrate Polymers* 265 (April). Elsevier Ltd: 118101. doi:10.1016/j.carbpol.2021.118101.
Fainerman, V. B., E. V. Aksenenko, J. Krägel, and R. Miller. 2016. "Thermodynamics, Interfacial Pressure Isotherms and Dilational Rheology of Mixed Protein-Surfactant Adsorption Layers." *Advances in Colloid and Interface Science* 233. Elsevier B. V.: 200–222. doi:10.1016/j.cis.2015.06.004.
Fainerman, V. B., E. H. Lucassen-Reynders, and R. Miller. 1998. "Adsorption of Surfactants and Proteins at Fluid Interfaces." *Colloids and Surfaces A: Physicochemical and Engineering Aspects* 143 (2–3): 141–65. doi:10.1016/S0927-7757(98)00585-8.
Flores, Zoila, Diego San-Martin, Tatiana Beldarraín-Iznaga, Javier Leiva-Vega, and Ricardo Villalobos-Carvajal. 2021. "Effect of Homogenization Method and Carvacrol Content on Microstructural and Physical Properties of Chitosan-Based Films." *Foods* 10 (1): 1–16. doi:10.3390/foods10010141.
Goodarzi, Fatemeh, and Sohrab Zendehboudi. 2019. "A Comprehensive Review on Emulsions and Emulsion Stability in Chemical and Energy Industries." *Canadian Journal of Chemical Engineering* 97 (1): 281–309. doi:10.1002/cjce.23336.
Haque, M. Minhaz Ul, Debora Puglia, Elena Fortunati, and Mariano Pracella. 2017. "Effect of Reactive Functionalization on Properties and Degradability of Poly(Lactic Acid)/Poly(Vinyl Acetate) Nanocomposites with Cellulose Nanocrystals." *Reactive and Functional Polymers* 110. Elsevier B. V.: 1–9. doi:10.1016/j.reactfunctpolym.2016.11.003.

Herrera-Rodríguez, S. E., R. J. López-Rivera, E. García-Márquez, M. Estarrón-Espinosa, and H. Espinosa-Andrews. 2019. "Mexican Oregano (*Lippia Graveolens*) Essential Oil-in-Water Emulsions: Impact of Emulsifier Type on the Antifungal Activity of Candida Albicans." *Food Science and Biotechnology* 28 (2): 441–48. doi:10.1007/s10068-018-0499-6.

Hussein, Jihan, Mona El-Bana, Eman Refaat, and Mehrez E. El-Naggar. 2017. "Synthesis of Carvacrol-Based Nanoemulsion for Treating Neurodegenerative Disorders in Experimental Diabetes." *Journal of Functional Foods* 37. Elsevier Ltd: 441–48. doi:10.1016/j.jff.2017.08.011.

Hussein, Jihan, Mona El-Banna, Khaled F. Mahmoud, Safaa Morsy, Yasmin Abdel Latif, Dalia Medhat, Eman Refaat, Abdel Razik Farrag, and Sherien M. El-Daly. 2017. "The Therapeutic Effect of Nano-Encapsulated and Nano-Emulsion Forms of Carvacrol on Experimental Liver Fibrosis." *Biomedicine and Pharmacotherapy* 90. Elsevier Masson SAS: 880–87. doi:10.1016/j.biopha.2017.04.020.

Jiang, Yang, Wang Dan, Li Feng, Dapeng Li, and Qingrong Huang. 2019. "Cinnamon Essential Oil Pickering Emulsion Stabilized by Zein-Pectin Composite Nanoparticles: Characterization, Antimicrobial Effect and Advantages in Storage Application." *International Journal of Biological Macromolecules*, no. xxxx. Elsevier B. V. doi:10.1016/j.ijbiomac.2019.10.103.

Kaptay, G. 2006. "On the Equation of the Maximum Capillary Pressure Induced by Solid Particles to Stabilize Emulsions and Foams and on the Emulsion Stability Diagrams." *Colloids and Surfaces A: Physicochemical and Engineering Aspects* 282–283: 387–401. doi:10.1016/j.colsurfa.2005.12.021.

Keivani Nahr, Fatemeh, Babak Ghanbarzadeh, Hossein Samadi Kafil, Hamed Hamishehkar, and Mohammadyar Hoseini. 2019. "The Colloidal and Release Properties of Cardamom Oil Encapsulated Nanostructured Lipid Carrier." *Journal of Dispersion Science and Technology* 0 (0). Taylor & Francis: 1–9. doi:10.1080/01932691.2019.1658597.

Khan, Imran, Ashutosh Bahuguna, Pradeep Kumar, Vivek K. Bajpai, and Sun Chul Kang. 2018. "In Vitro and in Vivo Antitumor Potential of Carvacrol Nanoemulsion against Human Lung Adenocarcinoma A549 Cells via Mitochondrial Mediated Apoptosis." *Scientific Reports* 8 (1). Springer US: 712–14. doi:10.1038/s41598-017-18644-9.

Kotsmar, Cs, V. Pradines, V. S. Alahverdjieva, E. V. Aksenenko, V. B. Fainerman, V. I. Kovalchuk, J. Krägel, M. E. Leser, B. A. Noskov, and R. Miller. 2009. "Thermodynamics, Adsorption Kinetics and Rheology of Mixed Protein-Surfactant Interfacial Layers." *Advances in Colloid and Interface Science* 150 (1). Elsevier B. V.: 41–54. doi:10.1016/j.cis.2009.05.002.

Kralova, Iva, and Johan Sjöblom. 2009. "Surfactants Used in Food Industry: A Review." *Journal of Dispersion Science and Technology* 30 (9): 1363–83. doi:10.1080/01932690902735561.

Kumar, Manish, Ram Singh Bishnoi, Ajay Kumar Shukla, and Chandra Prakash Jain. 2019. "Techniques for Formulation of Nanoemulsion Drug Delivery System: A Review." *Preventive Nutrition and Food Science* 24 (3): 225–34. doi:10.3746/pnf.2019.24.3.225.

Leclercq, Loïc, Jérémie Tessier, Grégory Douyère, Véronique Nardello-Rataj, and Andreea R. Schmitzer. 2020. "Phytochemical- And Cyclodextrin-Based Pickering Emulsions: Natural Potentiators of Antibacterial, Antifungal, and Antibiofilm Activity." *Langmuir* 36 (16): 4317–23. doi:10.1021/acs.langmuir.0c00314.

Low, Liang Ee, Sangeetaprivya P. Siva, Yong Kuen Ho, Eng Seng Chan, and Beng Ti Tey. 2020. "Recent Advances of Characterization Techniques for the Formation, Physical Properties and Stability of Pickering Emulsion." *Advances in Colloid and Interface Science* 277. Elsevier B. V: 102117. doi:10.1016/j.cis.2020.102117.

Mauriello, Eugenio, Giovanna Ferrari, and Francesco Donsì. 2021. "Effect of Formulation on Properties, Stability, Carvacrol Release and Antimicrobial Activity of Carvacrol Emulsions." *Colloids and Surfaces B: Biointerfaces* 197 (March 2020). Elsevier B. V.: 111424. doi:10.1016/j.colsurfb.2020.111424.

McClements, David Julian. 2007. "Critical Review of Techniques and Methodologies for Characterization of Emulsion Stability." *Critical Reviews in Food Science and Nutrition* 47 (7): 611–49. doi:10.1080/10408390701289292.

McClements, David Julian. 2012. "Nanoemulsions versus Microemulsions: Terminology, Differences, and Similarities." *Soft Matter* 8 (6): 1719–29. doi:10.1039/c2sm06903b.

McClements, David Julian. 2020. "Advances in Nanoparticle and Microparticle Delivery Systems for Increasing the Dispersibility, Stability, and Bioactivity of Phytochemicals." *Biotechnology Advances* 38 (May 2018). Elsevier: 107287. doi:10.1016/j.biotechadv.2018.08.004.

McClements, David Julian, and Cansu Ekin Gumus. 2016. "Natural Emulsifiers — Biosurfactants, Phospholipids, Biopolymers, and Colloidal Particles: Molecular and Physicochemical Basis of Functional Performance." *Advances in Colloid and Interface Science* 234. Elsevier B. V.: 3–26. doi:10.1016/j.cis.2016.03.002.

McClements, David Julian, and Seid Mahdi Jafari. 2018. "Improving Emulsion Formation, Stability and Performance Using Mixed Emulsifiers: A Review." *Advances in Colloid and Interface Science* 251. Elsevier B. V.: 55–79. doi:10.1016/j.cis.2017.12.001.

Mendes, J. F., L. B. Norcino, H. H. A. Martins, A. Manrich, C. G. Otoni, E. E. N. Carvalho, R. H. Piccoli, J. E. Oliveira, A. C. M. Pinheiro, and L. H. C. Mattoso. 2020. "Correlating Emulsion Characteristics with the Properties of Active Starch Films Loaded with Lemongrass Essential Oil." *Food Hydrocolloids* 100 (June 2019). Elsevier Ltd: 105428. doi:10.1016/j.foodhyd.2019.105428.

Moraes-Lovison, Marília, Luís F. P. Marostegan, Marina S. Peres, Isabela F. Menezes, Marluci Ghiraldi, Rodney A. F. Rodrigues, Andrezza M. Fernandes, and Samantha C. Pinho. 2017. "Nanoemulsions Encapsulating Oregano Essential Oil: Production, Stability, Antibacterial Activity and Incorporation in Chicken Pâté." *LWT - Food Science and Technology* 77. Elsevier Ltd: 233–40. doi:10.1016/j.lwt.2016.11.061.

Motta Felício, I., R. Limongi de Souza, C. de Oliveira Melo, K. Y. Gervázio Lima, U. Vasconcelos, R. Olímpio de Moura, and E. Eleamen Oliveira. 2021. "Development and Characterization of a Carvacrol Nanoemulsion and Evaluation of Its Antimicrobial Activity against Selected Food-Related Pathogens." *Letters in Applied Microbiology* 72 (3): 299–306. doi:10.1111/lam.13411.

Posocco, P., A. Perazzo, V. Preziosi, E. Laurini, S. Pricl, and S. Guido. 2016. "Interfacial Tension of Oil/Water Emulsions with Mixed Non-Ionic Surfactants: Comparison between Experiments and Molecular Simulations." *RSC Advances* 6 (6): 4723–29. doi:10.1039/c5ra24262b.

Prado, Karen S., Danielle Gonzales, and Márcia A. S. Spinacé. 2019. "Recycling of Viscose Yarn Waste through One-Step Extraction of Nanocellulose." *International Journal of Biological Macromolecules* 136. Elsevier B. V.: 729–37. doi:10.1016/j.ijbiomac.2019.06.124.

Rosen, Milton J, and Joy T Kunjappu. 2012. *Surfactants and Interfacial Phenomena. John Wiley & Sons*. 3rd ed. New Jersey: John Wiley & Sons. doi:10.1016/0166-6622(89)80030-7.

Safaya, M., and Y. C. Rotliwala. 2020. "Nanoemulsions: A Review on Low Energy Formulation Methods, Characterization, Applications and Optimization Technique." *Materials Today: Proceedings* 27. Elsevier Ltd.: 454–59. doi:10.1016/j.matpr.2019.11.267.

Scharamm, Laurier L. 2015. *Emulsions, Foams, Suspensions, and Aerosols*. 2nd ed. Vol. Microscien. Wiley-VCHVerlag GmbH& Co. http://repositorio.unan.edu.ni/2986/1/5624.pdf.

Shrestha, S., B. R. Wagle, A. Upadhyay, K. Arsi, D. J. Donoghue, and A. M. Donoghue. 2019. "Carvacrol Antimicrobial Wash Treatments Reduce Campylobacter Jejuni and Aerobic Bacteria on Broiler Chicken Skin." *Poultry Science* 98 (9). Published by Oxford University Press on behalf of Poultry Science Association 2019: 4073–83. doi:10.3382/ps/pez198.

Sow, Li Cheng, Felisa Tirtawinata, Hongshun Yang, Qingsong Shao, and Shifei Wang. 2017. "Carvacrol Nanoemulsion Combined with Acid Electrolysed Water to Inactivate Bacteria, Yeast in Vitro and Native Microflora on Shredded Cabbages." *Food Control* 76. Elsevier Ltd: 88–95. doi:10.1016/j.foodcont.2017.01.007.

Sufi-Maragheh, Parisa, Nasser Nikfarjam, Yulin Deng, and Nader Taheri-Qazvini. 2019. "Pickering Emulsion Stabilized by Amphiphilic PH-Sensitive Starch Nanoparticles as Therapeutic Containers." *Colloids and Surfaces B: Biointerfaces* 181 (December 2018): 244–51. doi:10.1016/j.colsurfb.2019.05.046.

Syed, Irshaan, Pratik Banerjee, and Preetam Sarkar. 2020. "Oil-in-Water Emulsions of Geraniol and Carvacrol Improve the Antibacterial Activity of These Compounds on Raw Goat Meat Surface during Extended Storage at 4°C." *Food Control* 107 (January): 106757. doi:10.1016/j.foodcont.2019.106757.

Walstra, Pieter. 1993. "Principles of Emulsion Formation." *Chemical Engineering Science* 48 (2): 333–49. doi:10.1016/0009-2509(93) 80021-H.

Wang, Pu, and Ying Wu. 2021. "A Review on Colloidal Delivery Vehicles Using Carvacrol as a Model Bioactive Compound." *Food Hydrocolloids* 120 (May). Elsevier Ltd: 106922. doi:10.1016/j.foodhyd.2021.106922.

Wang, Xiao Yan, and Marie Claude Heuzey. 2016. "Chitosan-Based Conventional and Pickering Emulsions with Long-Term Stability." *Langmuir* 32 (4): 929–36. doi:10.1021/acs.langmuir.5b03556.

Xiao, Jie, Yunqi Li, and Qingrong Huang. 2016. "Recent Advances on Food-Grade Particles Stabilized Pickering Emulsions: Fabrication, Characterization and Research Trends." *Trends in Food Science and Technology* 55. Elsevier Ltd: 48–60. doi:10.1016/j.tifs.2016.05.010.

Yang, Ruopeng, Jinyu Miao, Yuting Shen, Nan Cai, Chunpeng Wan, Liqiang Zou, Chuying Chen, and Jinyin Chen. 2021. "Antifungal Effect of Cinnamaldehyde, Eugenol and Carvacrol Nanoemulsion against Penicillium Digitatum and Application in Postharvest Preservation of Citrus Fruit." *LWT - Food Science and Technology* 141 (September 2020). Elsevier Ltd: 110924. doi:10.1016/j.lwt.2021.110924.

Zhang, Yefei, Vahid Karimkhani, Brian T. Makowski, Gamini Samaranayake, and Stuart J. Rowan. 2017. "Nanoemulsions and Nanolatexes Stabilized by Hydrophobically Functionalized Cellulose Nanocrystals." *Macromolecules* 50 (16): 6032–42. doi:10.1021/acs.macromol.7b00982.

Zhou, Yan, Shanshan Sun, Weiya Bei, Mohamed Reda Zahi, Qipeng Yuan, and Hao Liang. 2018. "Preparation and Antimicrobial Activity of Oregano Essential Oil Pickering Emulsion Stabilized by Cellulose Nanocrystals." *International Journal of Biological Macromolecules* 112. Elsevier B. V.: 7–13. doi:10.1016/j.ijbiomac.2018.01.102.

Zhu, Qiaomei, Yijun Pan, Xin Jia, Jinlong Li, Min Zhang, and Lijun Yin. 2019. "Review on the Stability Mechanism and Application of Water-in-Oil Emulsions Encapsulating Various Additives." *Comprehensive Reviews in Food Science and Food Safety* 18 (6): 1660–75. doi:10.1111/1541-4337.12482.

Zou, Tao, Mika H Sipponen, and Monika Österberg. 2019. "Natural Shape-Retaining Microcapsules With Shells Made of Chitosan-Coated Colloidal Lignin Particles." *Frontiers in Chemistry* 7 (May): 1–12. doi:10.3389/fchem.2019.00370.

Chapter 5

Carvacrol Encapsulation: Strategies, Preparation Methods, and Trends

Rafaela R. Ferreira and Derval S. Rosa[*]

Center for Engineering, Modeling, and Applied Social Sciences (CECS),
Federal University of ABC (UFABC),
Santo André, Brazil

Abstract

Essential oils (EO) have attracted significant attention as additives in chemical products (food) due to their olfactory, physical-chemical, and biological characteristics. The antimicrobial and antioxidant effects of oregano essential oil (OEO) are mainly attributed to the presence of carvacrol (phenolic monoterpene), which constitutes about 78-82% of the Origanum vulgare plant essential oil. However, essential oils (EO) are unstable compounds and susceptible to degradation when exposed to environmental stresses such as oxygen, temperature, and light. Thus, alternatives are sought to improve their application; in this sense, encapsulation is a method that offers a viable strategy to stabilize and prolong the EO release, allowing their application as an additive to food products. Encapsulation is defined as a method in which minuscule particles or droplets are surrounded by a cladding wall or embedded in a matrix. The matrix wall isolates the active compound as a functional barrier from the surrounding environment to prevent chemical and physical reactions and prolong their stability and bioavailability. Colloidal systems promote encapsulation, and several factors influence the effectiveness of

---

[*] Corresponding Author's E-mails: dervalrosa@yahoo.com.br or derval.rosa@ufabc.edu.br.

In: A Closer Look at Carvacrol
Editor: Zak A. Cunningham
ISBN: 978-1-68507-627-6
© 2022 Nova Science Publishers, Inc.

the process, such as the type of technique employed, emulsifier type and concentration, and wall material, reflecting on the encapsulation efficiency, particle size, and physical stability of encapsulated EOs. In addition, the associated benefits after encapsulation must be considered, that is, bioavailability, controlled release, and protection of the EO against environmental stresses. Thus, this chapter summarizes the recognized benefits, functional properties of various preparation and characterization methods, in which innovative manufacturing strategies and their mechanisms are demonstrated.

Keywords: carvacrol, encapsulation, characterization, kinetic modeling.

Introduction

Essential oils (EO) are naturally volatile aromatic liquids extracted from plants, such as flowers, buds, leaves, stems, and bark. A composition as defined physical properties of EO is greatly influenced by species, plant part, geographic origin, harvest time, development stage, age of plants, and extraction method (Pinto et al., 2020). Previous studies have revealed that, of these classes, a number of compounds have antioxidant and antibacterial activity. For this reason, OEs are an alternative to chemical preservatives and, therefore, are used to prepare safe foods with a positive impact on consumer health (Das et al., 2021; Dima et al., 2015).

The essential oil of oregano (OEO) has the active ingredient carvacrol (2-methyl-5-(1-methyl ethyl) phenol), which is a monoterpenoid compound formed by the linkage of two isoprene molecules with three functional group substituents (Figure 1) (Tao, Sedman, and Ismail 2021). It is a component of many aromatic plants of the Labiatae family that are commonly used as spices in cooking and for therapy/prevention purposes in folk medicine. The scientific interest in this molecule is due to its broad spectrum of biological and pharmacological properties, such as anti-inflammatory, antimicrobial, antioxidant, antitumor, anti-hepatotoxic, and insecticide activities (Scaffaro, Maio, and Nostro 2020).

Concerning antimicrobial activity, several investigations have demonstrated the efficacy of carvacrol against deteriorating pathogens or microorganisms, including clinically relevant drug-resistant and sessile lifestyle strains (Rajić et al., 2021; Tao, Sedman, and Ismail 2021). Carvacrol is classified as generally recognized as safe (GRAS) by the US Food and Drug Administration and included in the list of permitted additives for direct addition to foods for human consumption is approved by the European Commission for use as a food flavoring (Ayres Cacciatore et al., 2020). However, EOO is sensitive to oxygen, light, and high temperature. These factors contribute to the degradation of EO and, consequently, to reducing their biological potential. Therefore, encapsulation emerges as a strategy for antimicrobial protection against these previously discussed problems. In addition, encapsulation can provide more significant interaction with the surface and the possibility of controlled release of the active ingredient (Li et al., 2015).

Encapsulation-based delivery systems are a promising approach to overcome challenges containing free OEO (Kharat and McClements 2019). The main advantage of encapsulation systems lies in the fact that carvacrol can be developed through various preparation methods to improve its digestion and release properties and its inhibition property for pathogens. Therefore, this chapter will focus on discussing the effects of encapsulation vehicles and the recognized benefits and functional properties of various preparation and characterization methods. Kinetic modeling is concerned, which is a key factor in understanding the dynamic release control of encapsulated EOs. Furthermore, new trends in the development of EO capsules are also highlighted.

General Aspects of Encapsulation

Carvacrol ($C_{10}H_{14}O$) is a liquid phenolic monoterpenoid, 2-methyl-5-(1-methyl ethyl) phenol, present in the EO of oregano, thyme, peppers, wild bergamot, and other plants. (Gabrieli et al., 2020). Carvacrol has lipophilic properties and a density of 0.976 g / mL at room temperature (25 °C); it is insoluble in water but highly soluble in ethanol, acetone, and diethyl ether (Sharifi-Rad et al., 2018). As recently reviewed by several authors, this compound has a wide range of biological activities, including antimicrobials and antioxidants, as shown in Figure 1. However, carvacrol applications are limited due to their high volatility and low water solubility (Comunian and Favaro-Trindade 2016). Therefore, the successful encapsulation of carvacrol, lending it water solubility, long-term stability, and slow-release properties, has the potential to widely spread the application of this potent antimicrobial and antioxidant compound, extending shelf life, reducing spoilage, and better ensuring the safety of a variety of food products (Sun, Cameron, and Bai 2019).

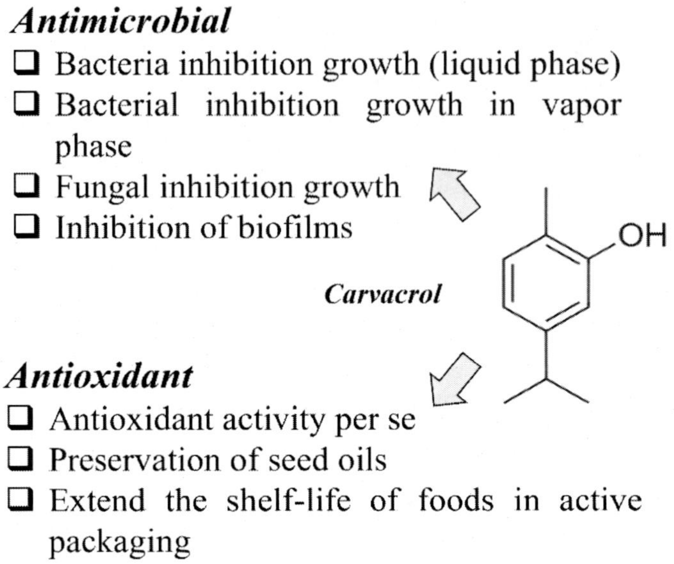

Antimicrobial
- Bacteria inhibition growth (liquid phase)
- Bacterial inhibition growth in vapor phase
- Fungal inhibition growth
- Inhibition of biofilms

Carvacrol

Antioxidant
- Antioxidant activity per se
- Preservation of seed oils
- Extend the shelf-life of foods in active packaging

Figure 1. The bioactivity of carvacrol focuses on its antimicrobial and antioxidant properties.

The encapsulation process involves coating or trapping solid, liquid, or gaseous particles in thin films using various active agents (Cheng et al., 2019; Dong et al., 2020; Shemesh et al., 2015). During encapsulation, the core materials are surrounded by a wall, which acts as a physical barrier to protect them from external factors. The particles obtained are called nanocapsules (nm) or microcapsules (μm). The sizes and shapes of the capsules depend on the materials and methods used to prepare them and mainly on the core materials and the shell deposition process (Eun et al., 2020). Different morphologies can be obtained, and the main objectives of the encapsulation process are presented below:

i. Reducing the transfer rate of the core material into the surrounding material;
ii. Protect core materials from undesirable environmental conditions;
iii. Prevent incompatibility and reactivity of compounds;
iv. Modify the physical characteristics of the original materials to facilitate handling;
v. Control the release of core materials;
vi. Offers better storage conditions, avoiding degradative reactions such as dehydration and oxidation.

A Comparison of Encapsulation Systems

Several studies have been carried out to develop carvacrol release systems to address the challenges associated with low water solubility and high chemical instability (Granata et al., 2018; Santos et al., 2019; Wang and Wu 2021). The nature of the encapsulating agents and the process used to produce the capsule influence the protection of the EO. To obtain them, there are physical methods (spray drying and lyophilization), physicochemical methods (coacervation, emulsification, liposome, and ionic gelling), and chemical methods (Interfacial polymerization and Complexation by molecular inclusion) (Hu and Luo 2021), the main feature of the different encapsulation systems is schematically represented in Figure 2. Furthermore, the functional performance of each of these delivery systems depends on their structural and physicochemical attributes, such as particle composition, particle size, morphology, stability, optical properties, rheology, and sensory characteristics.

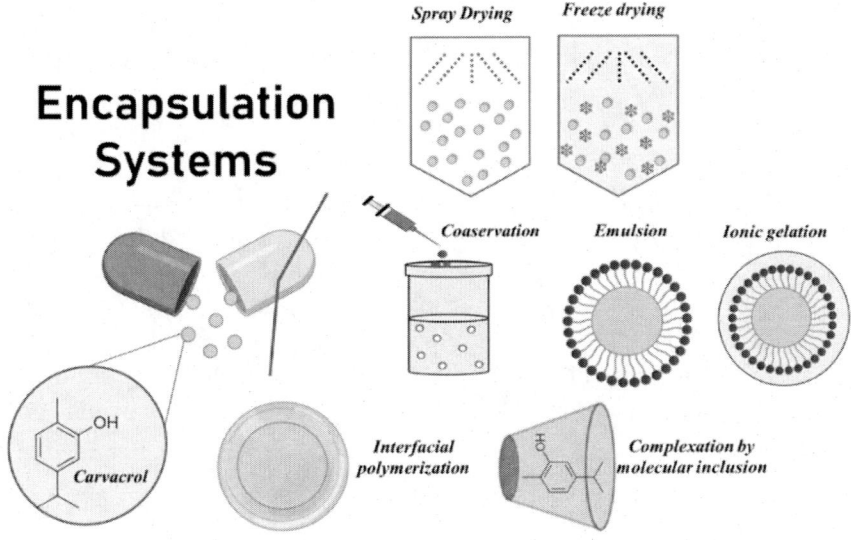

Figure 2. Schematic illustration of different carvacrol essential oil encapsulation systems using other methods.

Physical Methods

Spray Drying

Spray drying has been widely used on an industrial scale since the 1950s, mainly for fats, oils, flavors, and coloring (Francisco et al. 2020; Hernández-Nava et al., 2020; Mehran, Masoum, and Memarzadeh 2020; Radünz et al., 2020). Spray drying is highly suitable for heat-sensitive food ingredients. Several stages are involved in the encapsulation process using spray drying. Encapsulation efficiency depends on a number of factors such as feed flow rate, inlet/outlet air temperature, feed temperature, and wall materials (Bakry et al., 2016). The process can be conducted by changing the emulsion of the paste from a liquid form to a powder in a continuous operation procedure. The basic principle of this method is to dissolve the core/wall materials in water to prepare an emulsion in liquid form and then feed this emulsion into a hot medium (100–300 °C) to evaporate the water. The final dry product can be collected in powder or agglomerated particles, depending on the nature of the feed materials, dryer operation design, and operating conditions. The high temperature of the drying chamber facilitates the evaporation of water from the droplets (Mohammed et al., 2020). Advantages of spray drying include good reconstitution characteristics, low water activity, suitability for transport, good encapsulation efficiency, and good final product stability (Eun et al., 2020). However, the scarcity of good wall materials and the micro and nanocapsule powder agglomeration properties are the limitations of spray drying during the encapsulation process due to the damage that high temperatures can cause to sensitive bioactive compounds (Suganya and Anuradha 2017). To solve the problem of high temperatures, the effect of using reduced pressure and the absence of oxygen during the spray drying process (vacuum spray drying) was investigated by de Melo Ramos et al. (2019) (de Melo Ramos, Silveira Júnior, and Prata 2019).

Freeze Drying

Lyophilization, also known in the context of the encapsulation technique, is a multi-step process consisting of a first freezing step, followed by sublimation (primary drying) and subsequent desorption (secondary drying). The most significant advantage of freeze-drying is that it is a simple process at a low operating temperature with no air that produces superior and prolonged quality products, avoiding deterioration caused by oxidation or chemical modification (Razola-Díaz et al., 2021). Briefly, freeze-drying is the most suitable technique for dehydrating almost all heat-sensitive substances (El-Messery et al., 2020;

Rodriguez et al., 2019). This feature makes the freeze-drying process especially attractive for drying thermosensitive and bioactive components, such as essential oils, minimizing product damage caused by the high temperatures applied in the spray drying process (Chen et al., 2013). However, the freeze-drying technique has some drawbacks, such as the long process time (more than 20 hours), the high capital, and the high operating costs compared to other methods. In addition, the porous structure of freeze-dried powders, due to the sublimation of ice during the process, limits the control over particle size, as they must be crushed or converted into fine powders after drying (Ozkan et al., 2019).

Physicochemical Methods

Coaservation

The coacervation technique can be defined as a colloidal phenomenon involving the liquid-liquid phase separation of a single or a mixture of two polymers with opposite charges in an aqueous solution triggered by electrostatic interactions, hydrogen bonds, hydrophobic interactions, induced attractive interactions by polarization, as well as chemical or enzymatic crosslinking agents including glutaraldehyde, or transglutaminase (Ozkan et al., 2019). Coacervate formation requires close monitoring of different process parameters for complex coacervation, such as pH, temperature, ionic strength, agitation rate, biopolymer ratio, molecular weight, and biopolymer concentration (Eghbal and Choudhary 2018; Tang, Scher, and Jeoh 2020).

Despite the importance of biopolymer properties and coacervation process parameters, the thermodynamics of biopolymer bonding is the most important cause of coacervation. During complex coacervation, the electrostatic attraction of the polyelectrolyte favors the binding of the ligand involved in the interaction. Ligand binding leads to association, nucleation, and cluster formation, followed by coacervation. The binding mechanism to ligand and coacervates consists of the change in entropy and enthalpy, the number of binding sites, and the magnitude of the forces that control the polyelectrolyte interaction (Muhoza et al., 2020). Furthermore, coacervation is known for its simplicity, low cost, scalability, and reproducibility in the encapsulation of food ingredients, which results in high encapsulation efficiency, even with a very high payload (up to 99%) (Timilsena et al., 2019).

Emulsion

Emulsions are colloidal systems consisting of an immiscible liquid dispersed in another immiscible liquid in the form of droplets. Emulsions are considered water-in-oil when the dispersed phase is an aqueous solution. The continuous phase consists of an oil or lipid molecule, and oil-in-water consists of an oil phase dispersed in a continuous aqueous phase. To reduce the interfacial tension between the dispersed and continuous phases, forming a thin interfacial film of molecule surrounding each drop of oil, and to provide stable systems, an emulsifying agent or surfactant is required. The stabilized by a surfactant is a determining factor in the particle size range (Barradas and de Holanda e Silva 2021). Emulsions are mainly classified according to their particle diameter and thermodynamic stability; microemulsions have a smaller particle size (5–50 nm) and are more thermodynamically stable, while nanoemulsions have a droplet size of 20–200 nm and are thermodynamically unstable (Jin et al., 2016). These attractive natures have stimulated the preparation of EO nanoemulsions to preserve the product's appearance, increase bioavailability, reduce interaction and prevent oxidation (Basak and Guha 2018). Preparation methods can be distinguished between high energy (high pressure or mechanical homogenization and ultrasonication) and low energy (divided into isothermal and thermal processes) methods (Pavoni et al., 2019).

Ionic Gelation

Ionic gelling is one of the microencapsulation techniques based on the ability to crosslink polyelectrolytes in the presence of multivalent ions, such as Ca^{2+}, Ba^{2+}, and Al^{3+}. The interaction is a bivalent and trivalent ionic crosslinking between the ions of the divalent molecule and the envelope polymer (Ozkan et al., 2019). Ionic gelling can be performed by atomization, extrusion, co-extrusion, or electrostatic deposition processes (Bakry et al., 2016). Typically, a polymeric or hydrocolloid solution is dropped or atomized into an ionic solution under constant agitation. The active compound to be encapsulated is dissolved in the polymer solution. The drops that reach the ionic solution immediately form spherical gel structures, which contain the active dispersed throughout the polysaccharide matrix. It is a simple and easy procedure, does not require specialized equipment, high temperature, or organic solvent, and can be considered low cost (Ferreira, Souza, and Rosa 2021). However, one of its disadvantages is the heterogeneous gelation of gel particles due to the diffusion mechanism, as the surface gelation usually occurs before the core gelation, which becomes a softcore (Kurozawa and Hubinger 2017).

Chemical Methods

Interfacial Polymerization
Wall formation in this technique is characterized by polymerization, in which hydrophilic and lipophilic monomers interconnect in an oil-water emulsion and react to form a polymeric membrane on the surface of the drop or particle. The performance and quality of polymeric membranes manufactured using this technique can be optimized by controlling process parameters, including monomer concentrations, temperature, mixing speed, and reaction time. Mainly four polymers have been developed to produce microcapsules by interfacial polymerization, consisting of polyamides, polyurethanes, and polyesters (Razola-Díaz et al., 2021). The interfacial polymerization technique has potential advantages, including possible control of average capsule size and membrane thickness, high active compound loading, versatile and stable mechanical and chemical properties of the membrane, low cost, ease to scale up, simplicity and reliability of the process (Perignon et al., 2015). On the other hand, some factors limit the application of this technique, such as the difficulty of producing this oil-water interface (McClements 2020).

Complexation by Molecular Inclusion
Molecular inclusion is an encapsulation technique that occurs at the molecular level, consisting of the entrapment of the host (active) compound by a host (polymer) utilizing physicochemical forces, such as hydrogen bonds, van der Waals forces, or hydrophobic interactions. The most common "host" molecules are cyclodextrins, composed of a hydrophilic outer part and an inner hydrophobic part. The nonpolar guest molecule can be trapped in the internal apolar cavity through hydrophobic interactions (Ozkan et al., 2019). The formation of complexes with a hydrophobic molecule is because the displacement of water from within the cyclodextrin is energetically favorable. Complexation increases the solubility of hydrophobic host molecules in water, and the ingredient is protected from degradation (McClements 2020). This technique is worthwhile for the slow release of volatile compounds, more excellent stability when compared to other methods.

Characterization of Nano or Microcapsules

Nano and microcapsules can be characterized in two instances of the process using various forms of analysis: a) before the drying process, where measurements of zeta potential can examine the charge of the colloidal particles in the dispersion, viscosity, and viscoelastic properties determined by rheological analysis and b) after the drying process, nano and microcapsule powders can be characterized for their structural, antimicrobial and physicochemical properties using different characterization techniques, such as Fourier Transform Infrared Spectroscopy (FTIR), analysis thermogravimetry (TGA), scanning electron microscopy (SEM) and transmission electron microscopy (TEM) (Tavares, Santos, and Zapata Noreña 2021). Both are shown in Figure 3.

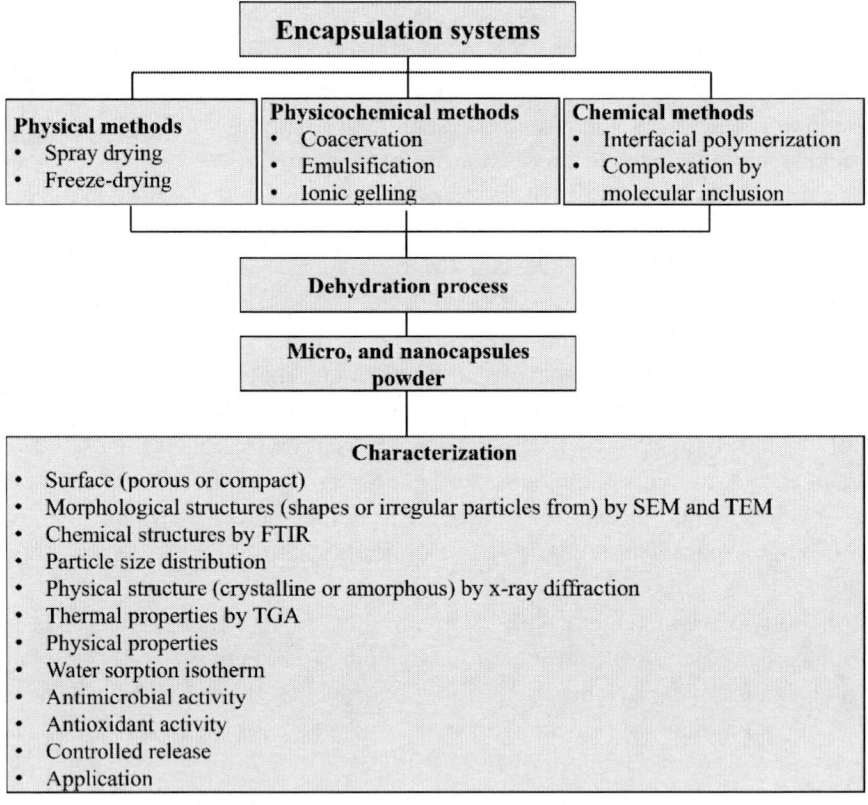

Figure 3. Encapsulation techniques and analytical techniques used in the characterization of nano and microcapsules after the drying process.

Carvacrol Kinetics

The carvacrol release kinetics is in accordance with the differences in solvent polarity and the corresponding chemical affinity between the polymer matrix (Ramos et al., 2012). The relaxation of the polymer in contact with the solvents was not coupled with the diffusion of the compound, and three steps can be taken for the release process: (a) the diffusion of the solvent into the polymer matrix, (b) the relaxation of the in-line network with solvation and plasticization, and (c) diffusion of the compound through the relaxed polymer network until thermodynamic equilibrium between the phases (polymer/food system) is reached. On balance, the compound's affinity to the solvated polymer and its solubility in the liquid food system will determine the partition coefficient for the compound administered. The diffusion of the compound through the matrix will be affected by the solvent impregnation in the polymeric network and by the interactions established between the components (Requena, Vargas, and Chiralt 2017). These steps can be coupled, mainly 2 and 3, depending on the characteristic relaxation time of the polymers and the diffusion times of the diffusion compound, giving rise to an anomalous transport behavior (Siepmann and Peppas 2011).

Furthermore, the physicochemical properties of the active compounds, as well as the wall materials and the core: wall material ratio, influence the release of active compounds from the formulation and, therefore, modify the release kinetics. The mathematical models of drug release with greater application and which best describe drug release phenomena are, in general, the Higuchi model, the zero-order model, the first-order model, and the Korsmeyer-Peppas model (Samaha, Shehayeb, and Kyriacos 2009). Thus, different kinetic models can be used for this analysis. The linear regression correlation coefficient (R^2) value indicates the best fit of the different models described in Table 1.

Rajić et al. (2021) used the anti-solvent precipitation technique to manufacture composite nanoparticles (NPs) of zein/rosin loaded with carvacrol (Z/R) and zein/shellac (Z/S). Cumulative release curves were fitted using the Higuchi release model; the results showed that resin type and participation affect encapsulation efficiency and carvacrol release. It was also demonstrated that the addition of resins improved the release of nanoparticles compared to single zein nanoparticles (Rajić et al., 2021).

Table 1. Summary of mathematical models used to study the release profiles of encapsulated carvacrol essential oil

Mathematical Model	Equation	Parameters	References
Higuchi	$\dfrac{M_t}{M_\infty} = k_H\, t^{1/2}$	Where M_t is the amount of contaminant released at any given time, t is the time of release, M_∞ Is the total amount of contaminant released, and kH is Higuchi's diffusion constant.	(Argin, Kofinas, and Lo 2014; Simoni et al., 2017)
Korsmeyer-Peppas	$\dfrac{M_t}{M_\infty} = K \cdot t^n$	Mt, and M∞ represent as masses, at time t and equilibrium, respectively, k is a constant, and n is a parameter that describes the diffusion mechanism	(Marciano et al., 2021)
Zero-order release kinetics	$Q = Q_0 + k_0 t$	Q is the amount of drug released or dissolved, Q0 is the initial drug in solution (usually zero), and K0 is the zero-order release constant.	(Paarakh et al., 2019)
First-order release kinetics	$\log C_T = \log C_0 - \dfrac{kT}{2.303}$	C0 is the initial concentration of the drug, and CT is the concentration of hydrogen in solution at time t.	(Paarakh et al., 2019)

Keawchaoon and Yoksan (2011) developed chitosan nanoparticles loaded with carvacrol by a two-step method, i.e., oil-in-water emulsion and ionic gelling of chitosan with chitosan tripolyphosphate. The release mechanism and release kinetics of carvacrol from chitosan nanoparticles was also investigated using the Korsmeyer-Peppas model; the release of carvacrol from chitosan nanoparticles reached plateau level on day 30, with release amounts of 53% in acetate buffer solution at pH 3 and 23% and 33% in phosphate buffer solution at pH 7 and 11, respectively. The release mechanism followed a Fickian behavior. The release rate was higher in acidic medium to alkaline or neutral medium, respectively (Keawchaoon and Yoksan 2011).

Baranauskaite, Kopustinskiene, and Bernatoniene (2019) will develop the composition of the wall material (gelatin supplemented with gum arabic, Tween 20, and β-cyclodextrin) of EO carvacrol microcapsules prepared by spray drying to maximize product yield and efficiency of encapsulation. The authors investigated the four models proposed in Table 1 and reported that if the release of the active compound follows zero-order kinetics, the release rate is independent of the concentration of the active compound and depends on it if the release follows a kinetic first order. The Higuchi equation suggests the release of the active compound by diffusion, while the Korsmeyer-Peppas power law equation defines the type of diffusion. So the stability investigations of the prepared microcapsules revealed that their physicochemical properties and rosmarinic acid and carvacrol content remained unchanged for 12 months (Baranauskaite, Kopustinskiene, and Bernatoniene 2019).

Furthermore, the effectiveness of encapsulation loaded with antimicrobial compounds depends not only on the nature of the active compounds but also on the ability to release an adequate concentration for a given application; thus, an equilibrium contact time (partition coefficient) is expected. This, in turn, depends on the interactions of the action with the polymeric matrix and its solubility in the investigated system. Therefore, the release kinetics of the active compound in food over storage time is crucial in ensuring antimicrobial efficacy and food safety (Fernández-Pan et al., 2015). In this sense, several mathematical models, commonly found in the literature, have been used to determine the release rate of active compounds and guarantee the minimum inhibitory concentration (MIC), ensuring an effective concentration in the system and certifying food safety.

Conclusion

The effects of encapsulation techniques on active agents (EO carvacrol) by different physical, physicochemical, and chemical methods were discussed, along with each method's advantages, disadvantages, and potential applications. Based on the encapsulation mechanisms, we conclude that the protection of bioactive compounds to be encapsulated can be improved when using carrier agents. The thermal stability of the polymer matrix is also effective in core material bioactivities against detrimental conditions. The parameters related to the physical-chemical functions of the encapsulated material must be optimized for each encapsulation technique, core, and wall material; thus, it allows to obtain narrower size distributions, which avoids significant product losses and allows for greater nutritional value. The delivery efficiency and release profile of carvacrol depend on the corresponding ingredients and encapsulation techniques. Most research indicates that EO carvacrol is a good topic for the food, cosmetics, and packaging industries. Besides several new technologies that have been used to produce functional EO, the use of other sustainable technologies. The information on EO encapsulation provided in this review may encourage further research into using this ingredient for different scopes, improving the range of applications. Thus, the application in active packaging and the study of controlled release can be very interesting research areas to be developed.

References

Argin, Sanem, Peter Kofinas, and Y. Martin Lo. 2014. "The Cell Release Kinetics and the Swelling Behavior of Physically Crosslinked Xanthan-Chitosan Hydrogels in Simulated Gastrointestinal Conditions." *Food Hydrocolloids* 40: 138–44.

Ayres Cacciatore, Fabíola et al., 2020. "Carvacrol Encapsulation into Nanostructures: Characterization and Antimicrobial Activity against Foodborne Pathogens Adhered to Stainless Steel." *Food Research International* 133(December 2019): 109143. https://doi.org/10.1016/j.foodres.2020.109143.

Bakry, Amr M. et al., 2016. "Microencapsulation of Oils: A Comprehensive Review of Benefits, Techniques, and Applications." *Comprehensive Reviews in Food Science and Food Safety* 15(1): 143–82.

Baranauskaite, Juste, Dalia M. Kopustinskiene, and Jurga Bernatoniene. 2019. "Impact of Gelatin Supplemented with Gum Arabic, Tween 20, and β-Cyclodextrin on the Microencapsulation of Turkish Oregano Extract." *Molecules* 24(1).

Barradas, Thaís Nogueira, and Kattya Gyselle de Holanda e Silva. 2021. "Nanoemulsions of Essential Oils to Improve Solubility, Stability and Permeability: A Review." *Environmental Chemistry Letters* 19(2): 1153–71. https://doi.org/10.1007/s10311-020-01142-2.

Basak, Suradeep, and Proshanta Guha. 2018. "A Review on Antifungal Activity and Mode of Action of Essential Oils and Their Delivery as Nano-Sized Oil Droplets in Food System." *Journal of Food Science and Technology* 55(12): 4701–10. https://doi.org/10.1007/s13197-018-3394-5.

Chen, Qiong et al., 2013. "Properties and Stability of Spray-Dried and Freeze-Dried Microcapsules Co-Encapsulated with Fish Oil, Phytosterol Esters, and Limonene." *Drying Technology* 31(6): 707–16.

Cheng, Meng et al., 2019. "Characterization and Application of the Microencapsulated Carvacrol/Sodium Alginate Films as Food Packaging Materials." *International Journal of Biological Macromolecules* 141: 259–67. https://doi.org/10.1016/j.ijbiomac. 2019.08.215.

Comunian, Talita A., and Carmen S. Favaro-Trindade. 2016. "Microencapsulation Using Biopolymers as an Alternative to Produce Food Enhanced with Phytosterols and Omega-3 Fatty Acids: A Review." *Food Hydrocolloids* 61: 442–57. http://dx.doi.org/10.1016/ j.foodhyd.2016.06.003.

Das, Somenath et al., 2021. "Eugenol Loaded Chitosan Nanoemulsion for Food Protection and Inhibition of Aflatoxin B1 Synthesizing Genes Based on Molecular Docking." *Carbohydrate Polymers* 255(September 2020): 117339. https://doi.org/10.1016/j.carbpol. 2020.117339.

de Melo Ramos, Fernanda, Vivaldo Silveira Júnior, and Ana Silvia Prata. 2019. "Assessing the Vacuum Spray Drying Effects on the Properties of Orange Essential Oil Microparticles." *Food and Bioprocess Technology* 12(11): 1917–27.

Dima, Cristian, Livia Pa, Alina Cantaragiu, and Petru Alexe. 2015. *The Kinetics of the Swelling Process and the Release Mechanisms of Coriandrum Sativum L . Essential Oil from Chitosan / Alginate / Inulin Microcapsules.*

Dong, Hao, Jiapeng He, Kaijun Xiao, and Chao Li. 2020. "Temperature-Sensitive Polyurethane (TSPU) Film Incorporated with Carvacrol and Cinnamyl Aldehyde: Antimicrobial Activity, Sustained Release Kinetics and Potential Use as Food Packaging for Cantonese-Style Moon Cake." *International Journal of Food Science and Technology* 55(1): 293–302.

Eghbal, Noushin, and Ruplal Choudhary. 2018. "Complex Coacervation: Encapsulation and Controlled Release of Active Agents in Food Systems." *LWT - Food Science and Technology* 90(May 2017): 254–64. https://doi.org/10.1016/j.lwt.2017.12.036.

El-Messery, Tamer Mohammed, Umit Altuntas, Gokce Altin, and Beraat Özçelik. 2020. "The Effect of Spray-Drying and Freeze-Drying on Encapsulation Efficiency, in Vitro Bioaccessibility and Oxidative Stability of Krill Oil Nanoemulsion System." *Food Hydrocolloids* 106(October 2019).

Eun, Jong Bang, Ahmed Maruf, Protiva Rani Das, and Seung Hee Nam. 2020. "A Review of Encapsulation of Carotenoids Using Spray Drying and Freeze Drying." *Critical Reviews in Food Science and Nutrition* 60(21): 3547–72. https://doi.org/10.1080/10408398.2019.1698511.

Fernández-Pan, I., J.I. Maté, C. Gardrat, and V. Coma. 2015. "Effect of Chitosan Molecular Weight on the Antimicrobial Activity and Release Rate of

Carvacrol-Enriched Films." *Food Hydrocolloids* 51: 60–68. https://linkinghub.elsevier.com/retrieve/pii/S0268005X15001988.

Ferreira, Rafaela R., Alana G. Souza, and Derval S. Rosa. 2021. "Essential Oil-Loaded Nanocapsules and Their Application on PBAT Biodegradable Films." *Journal of Molecular Liquids* 337: 116488.

Francisco, Cristhian Rafael Lopes et al., 2020. "Plant Proteins at Low Concentrations as Natural Emulsifiers for an Effective Orange Essential Oil Microencapsulation by Spray Drying." *Colloids and Surfaces A: Physicochemical and Engineering Aspects* 607(May): 125470. https://doi.org/10.1016/j.colsurfa.2020.125470.

Gabrieli, Alana et al., 2020. "International Journal of Biological Macromolecules Synergic Antimicrobial Properties of Carvacrol Essential Oil and Montmorillonite in Biodegradable Starch Fi Lms." *International Journal of Biological Macromolecules* 164: 1737–47. https://doi.org/10.1016/j.ijbiomac.2020.07.226.

Granata, Giuseppe et al., 2018. "Essential Oils Encapsulated in Polymer-Based Nanocapsules as Potential Candidates for Application in Food Preservation." *Food Chemistry* 269(March): 286–92.

Hernández-Nava, Ruth et al., 2020. "Encapsulation of Oregano Essential Oil (Origanum Vulgare) by Complex Coacervation between Gelatin and Chia Mucilage and Its Properties after Spray Drying." *Food Hydrocolloids* 109(January).

Hu, Qiaobin, and Yangchao Luo. 2021. "Chitosan-Based Nanocarriers for Encapsulation and Delivery of Curcumin: A Review." *International Journal of Biological Macromolecules* 179: 125–35. https://doi.org/10.1016/j.ijbiomac.2021.02.216.

Jin, Weiping et al., 2016. Emulsions *Nanoemulsions for Food: Properties, Production, Characterization, and Applications*. Elsevier Inc. http://dx.doi.org/10.1016/B978-0-12-804306-6/00001-5.

Keawchaoon, Lalita, and Rangrong Yoksan. 2011. "Preparation, Characterization and in Vitro Release Study of Carvacrol-Loaded Chitosan Nanoparticles." *Colloids and Surfaces B: Biointerfaces* 84(1): 163–71. http://dx.doi.org/10.1016/j.colsurfb.2010.12.031.

Kharat, Mahesh, and David Julian McClements. 2019. "Recent Advances in Colloidal Delivery Systems for Nutraceuticals: A Case Study – Delivery by Design of Curcumin." *Journal of Colloid and Interface Science* 557: 506–18. https://doi.org/10.1016/j.jcis.2019.09.045.

Kurozawa, Louise Emy, and Miriam Dupas Hubinger. 2017. "Hydrophilic Food Compounds Encapsulation by Ionic Gelation." *Current Opinion in Food Science* 15: 50–55. http://dx.doi.org/10.1016/j.cofs.2017.06.004.

Li, Wei et al., 2015. "Influence of Surfactant and Oil Composition on the Stability and Antibacterial Activity of Eugenol Nanoemulsions." *Lwt* 62(1): 39–47. http://dx.doi.org/10.1016/j.lwt.2015.01.012.

Marciano, Jéssica S. et al., 2021. "Biodegradable Gelatin Composite Hydrogels Filled with Cellulose for Chromium (VI) Adsorption from Contaminated Water." *International Journal of Biological Macromolecules* 181: 112–24.

McClements, David Julian. 2020. "Advances in Nanoparticle and Microparticle Delivery Systems for Increasing the Dispersibility, Stability, and Bioactivity of Phytochemicals." *Biotechnology Advances* 38(August 2018): 107287. https://doi.org/10.1016/j. biotechadv.2018.08.004.

Mehran, Mehdi, Saeed Masoum, and Mohammadreza Memarzadeh. 2020. "Microencapsulation of Mentha Spicata Essential Oil by Spray Drying: Optimization, Characterization, Release Kinetics of Essential Oil from Microcapsules in Food Models." *Industrial Crops and Products* 154(June): 112694. https://doi.org/10.1016/j.indcrop. 2020.112694.

Mohammed, Nameer Khairullah et al., 2020. "Spray Drying for the Encapsulation of Oils—A Review." *Molecules* 25(17): 1–16.

Muhoza, Bertrand et al., 2020. "Microencapsulation of Essential Oils by Complex Coacervation Method: Preparation, Thermal Stability, Release Properties and Applications." *Critical Reviews in Food Science and Nutrition* 0(0): 1–20. https://doi.org/10.1080/10408398. 2020.1843132.

Ozkan, Gulay et al., 2019. "A Review of Microencapsulation Methods for Food Antioxidants: Principles, Advantages, Drawbacks and Applications." *Food Chemistry* 272(August 2018): 494–506. https://doi.org/10.1016/j.foodchem.2018.07.205.

Paarakh, M Padmaa, Preethy A N I Jose, C M Setty, and G V Peter. 2019. "Release Kinetics – Concepts and Applications." *International Journal of Pharmacy Research & Technology* 8(1): 12–20.

Pavoni, Lucia et al. 2019. "Green Micro-and Nanoemulsions for Managing Parasites, Vectors and Pests." *Nanomaterials* 9(9).

Perignon, Carole et al., 2015. "Microencapsulation by Interfacial Polymerisation: Membrane Formation and Structure." *Journal of Microencapsulation* 32(1): 1–15.

Pinto, Lívia et al., 2020. *Food Hydrocolloids Encapsulation of Black Pepper (Piper Nigrum L .) Essential Oil with Gelatin and Sodium Alginate by Complex Coacervation.* 102(September 2019).

Radünz, Marjana et al., 2020. "Antimicrobial Potential of Spray Drying Encapsulated Thyme (*Thymus Vulgaris*) Essential Oil on the Conservation of Hamburger-like Meat Products." *International Journal of Food Microbiology* 330(January): 108696. https://doi.org/ 10.1016/j.ijfoodmicro.2020.108696.

Rajić, Danijela et al., 2021. "Zein–Resin Composite Nanoparticles with Coencapsulated Carvacrol." *Journal of Food Processing and Preservation* (March): 1–10.

Ramos, Marina, Alfonso Jiménez, Mercedes Peltzer, and María C. Garrigós. 2012. "Characterization and Antimicrobial Activity Studies of Polypropylene Films with Carvacrol and Thymol for Active Packaging." *Journal of Food Engineering* 109(3): 513–19. http://dx.doi.org/10.1016/j.jfoodeng.2011.10.031.

Razola-Díaz, María del Carmen, Eduardo Jesús Guerra-Hernández, Belén García-Villanova, and Vito Verardo. 2021. "Recent Developments in Extraction and Encapsulation Techniques of Orange Essential Oil." *Food Chemistry* 354(February).

Requena, Raquel, María Vargas, and Amparo Chiralt. 2017. "Release Kinetics of Carvacrol and Eugenol from Poly(Hydroxybutyrate-Co-Hydroxyvalerate) (PHBV) Films for Food Packaging Applications." *European Polymer Journal* 92(April): 185–93. http://dx.doi.org/10. 1016/j.eurpolymj.2017.05.008.

Rodriguez, Erica S. et al., 2019. "Effect of Natural Antioxidants on the Physicochemical Properties and Stability of Freeze-Dried Microencapsulated Chia Seed Oil." *Journal of the Science of Food and Agriculture* 99(4): 1682–90.

Samaha, D., R. Shehayeb, and S. Kyriacos. 2009. "Modeling and Comparison of Dissolution Profiles of Diltiazem Modified-Release Formulations." *Dissolution Technologies* 16(2): 41–46.

Santos, Jéssica D. C. et al., 2019. "Chemical Composition and Antimicrobial Activity of Satureja Montana Byproducts Essential Oils." *Industrial Crops and Products* 137(February): 541–48. https://doi.org/10.1016/j.indcrop.2019.05.058.

Scaffaro, Roberto, Andrea Maio, and Antonia Nostro. 2020. "Poly(Lactic Acid)/Carvacrol-Based Materials: Preparation, Physicochemical Properties, and Antimicrobial Activity." *Applied Microbiology and Biotechnology* 104(5): 1823–35.

Sharifi-Rad, Mehdi et al., 2018. "Carvacrol and Human Health: A Comprehensive Review." *Phytotherapy Research* 32(9): 1675–87.

Shemesh, Rotem et al., 2015. "Antibacterial and Antifungal LDPE Films for Active Packaging." *Polymers for Advanced Technologies* 26(1): 110–16.

Siepmann, Juergen, and Nicholas A. Peppas. 2011. "Higuchi Equation: Derivation, Applications, Use and Misuse." *International Journal of Pharmaceutics* 418(1): 6–12. http://dx.doi.org/10.1016/j.ijpharm.2011.03.051.

Simoni, Rayssa C. et al., 2017. "Effect of Drying Method on Mechanical, Thermal and Water Absorption Properties of Enzymatically Crosslinked Gelatin Hydrogels." *Anais da Academia Brasileira de Ciencias* 89(1): 745–55.

Suganya, V., and V. Anuradha. 2017. "Microencapsulation and Nanoencapsulation: A Review." *International Journal of Pharmaceutical and Clinical Research* 9(3): 233–39.

Sun, Xiuxiu, Randall G. Cameron, and Jinhe Bai. 2019. "Microencapsulation and Antimicrobial Activity of Carvacrol in a Pectin-Alginate Matrix." *Food Hydrocolloids* 92(January): 69–73. https://doi.org/10.1016/j.foodhyd.2019.01.006.

Tang, Yuting, Herbert B. Scher, and Tina Jeoh. 2020. "Industrially Scalable Complex Coacervation Process to Microencapsulate Food Ingredients." *Innovative Food Science and Emerging Technologies* 59(November 2019): 102257. https://doi.org/10.1016/j.ifset.2019.102257.

Tao, Ran, Jacqueline Sedman, and Ashraf Ismail. 2021. "Characterization and In Vitro Antimicrobial Study of Soy Protein Isolate Films Incorporating Carvacrol." *Food Hydrocolloids*: 107091. https://doi.org/10.1016/j.foodhyd.2021.107091.

Tavares, Loleny, Lúcia Santos, and Caciano Pelayo Zapata Noreña. 2021. "Bioactive Compounds of Garlic: A Comprehensive Review of Encapsulation Technologies, Characterization of the Encapsulated Garlic Compounds and Their Industrial Applicability." *Trends in Food Science and Technology* 114(March): 232–44.

Timilsena, Yakindra Prasad et al., 2019. "Complex Coacervation: Principles, Mechanisms and Applications in Microencapsulation." *International Journal of Biological Macromolecules* 121: 1276–86. https://doi.org/10.1016/j.ijbiomac.2018.10.144.

Wang, Pu, and Ying Wu. 2021. "A Review on Colloidal Delivery Vehicles Using Carvacrol as a Model Bioactive Compound." *Food Hydrocolloids* 120(May): 106922.

Index

A
acid, 18, 21, 31, 61, 76, 95, 101, 111, 161
Acitenobacter, 106
activation energy, 137
active compound, x, 42, 75, 78, 128, 129, 131, 149, 156, 157, 159, 161
acute lung injury, 22, 38, 46
acute myeloid leukemia, 29
acute promyelocytic leukemia, 28
acute respiratory distress syndrome, 23
additives, vii, ix, 1, 64, 75, 78, 86, 128, 149, 150
adhesion, 21, 23, 36, 60, 61, 91, 93, 95, 100, 110, 125
adsorption, 65, 131, 135, 136, 137, 142
angiotensin converting enzyme, 33
antibiotic, 9, 29, 30, 55, 56, 57, 59, 60, 81, 114, 128
antibiotic resistance, 57, 60, 81, 114
anti-cancer, ix, 24, 25, 28, 33, 127, 140, 142
anticancer activity, 48
antidepressant, 140, 142
antimicrobial, vii, viii, ix, 2, 4, 5, 9, 11, 12, 13, 15, 16, 17, 30, 32, 33, 36, 38, 40, 41, 42, 43, 44, 46, 47, 48, 49, 53, 54, 55, 56, 57, 58, 59, 61, 62, 63, 64, 65, 66, 69, 73, 74, 75, 76, 78, 79, 81, 84, 85, 86, 87, 90, 91, 101, 106, 107, 109, 110, 114, 115, 116, 118, 120, 121, 122, 128, 129, 130, 131, 133, 134, 135, 140, 141, 142, 144, 145, 146, 147, 149, 150, 151, 152, 158, 161, 162, 163, 164, 165, 166, 167
antioxidant, viii, ix, 2, 4, 5, 17, 18, 19, 20, 23, 24, 26, 28, 30, 32, 33, 34, 35, 37, 39, 40, 41, 42, 44, 46, 47, 48, 49, 50, 54, 76, 77, 78, 79, 80, 85, 86, 117, 119, 120, 122, 127, 140, 141, 142, 149, 150, 151, 152
apoptosis, 19, 24, 26, 27, 28, 29, 33, 36, 38, 41, 42, 43, 44, 49
attachment, 12, 13, 94, 95, 99, 123

B
Bacillus, 4, 68, 92, 94, 98, 100, 107, 109, 130
bacteria, 9, 10, 11, 14, 15, 16, 33, 38, 54, 56, 57, 58, 59, 60, 61, 63, 65, 70, 71, 76, 81, 85, 90, 91, 92, 93, 95, 97, 99, 100, 101, 106, 109, 110, 111, 112, 117, 118, 119, 122, 123, 124, 125, 131, 141
bacterial cells, 30, 65, 90
bacterial infection, 11, 14, 29, 46, 97
bacterial pathogens, 90
bacterial persistence, 109
bcl-2, 19, 26, 28
beef, 56, 62, 64, 65, 67, 69, 70, 71, 72, 73, 79, 82, 83, 84, 85, 87, 98, 99, 124
benefits, x, 30, 61, 62, 77, 150, 151
bioavailability, x, 5, 28, 31, 32, 33, 128, 130, 140, 149, 156
biofilms, viii, 10, 13, 16, 32, 37, 41, 45, 49, 62, 65, 76, 89, 90, 92, 93, 95, 96, 97, 98, 99, 100, 101, 109, 110, 111, 114, 115, 116, 119, 120, 122, 124, 125
biological activities, 2, 9, 44, 101, 151
biological activity, 37, 41, 103, 116, 119, 121
biologically active compounds, 2

biomedical applications, 140
biosynthesis, 15, 34, 43, 61, 102
biotechnology, 122, 124
blood, 20, 98, 133
blood flow, 21
blood vessels, 20
brain, 18, 30, 48, 140
breast cancer, viii, 2, 27, 35, 44
bronchial asthma, 23

C

cancer, 10, 17, 23, 24, 25, 27, 28, 29, 32, 33, 34, 38, 39, 40, 44, 47
cancer cells, 25, 27, 34
cancer progression, 28
carvacrol release kinetics, 159
causing food contamination, 92
cell cycle, 26, 27, 41, 44
cell death, 17, 36, 44, 48, 109
cell line, 12, 21, 26, 27, 29, 35, 51
changing environment, 130
characterization, x, 42, 48, 49, 85, 103, 115, 121, 124, 125, 143, 144, 145, 146, 147, 150, 151, 157, 158, 162, 163, 164, 165, 166, 167
chemical, ix, 3, 5, 8, 20, 30, 31, 33, 36, 41, 48, 49, 50, 54, 55, 56, 61, 64, 71, 78, 101, 102, 120, 122, 127, 132, 140, 142, 149, 150, 153, 154, 155, 157, 159, 161
chemical characteristics, ix, 64, 128, 140
chemical interaction, 142
chemical properties, 71, 101, 102, 157
chicken, 37, 49, 60, 62, 63, 69, 70, 74, 75, 76, 77, 80, 82, 83, 84, 92, 116
chitosan, 15, 64, 73, 131, 133, 160
classification, 98, 129, 130, 131
coaservation, 155
coatings, 48, 65, 73, 77, 78, 82, 113, 122
combines moisture, viii, 89
complexation by molecular inclusion, 153
composition, 5, 9, 12, 35, 41, 44, 46, 47, 48, 64, 68, 79, 99, 103, 115, 116, 117, 119, 120, 121, 122, 124, 132, 133, 134, 150, 153, 161

constituents, 2, 5, 9, 11, 15, 17, 44, 80, 81, 98, 116
consumers, viii, 55, 56, 66, 67, 77, 78, 90, 91
consumption, 12, 55, 67, 68, 69, 82, 92, 95, 96, 112, 116, 151
contaminated food, 12, 92, 107, 108
contamination, 12, 56, 57, 60, 64, 73, 90, 92, 93, 94, 95, 97, 99, 101, 106, 107, 114, 116, 118, 120, 121, 122, 141
cooking, 62, 67, 68, 69, 70, 71, 78, 79, 84, 85, 92, 94, 150
cosmetics, vii, 1, 2, 3, 9, 106, 140, 162
cytokines, 20, 21, 22, 23, 31, 44, 47, 141
cytotoxicity, 30, 32, 34, 35, 131

D

degradation, x, 62, 149, 151, 157
depth, vii, ix, 75, 128, 135, 142
derivatives, 30, 35, 39, 44, 54, 75, 117, 120
detection, 9, 51, 92, 113, 124, 140
diabetes, 17, 19, 37, 140
diffusion, 129, 137, 140, 156, 159, 160, 161
diseases, 2, 4, 9, 11, 13, 15, 21, 33, 50, 51, 58, 60, 90, 92, 99, 114, 116, 141
disinfection, 90, 92, 99, 109, 111
disorders, 17, 42, 140, 144
dispersion, ix, 93, 127, 132, 136, 137, 158
distillation, 5, 6, 7, 48, 49, 103, 104, 105, 122, 123
distribution, 62, 66, 125, 130, 138
DNA, 4, 19, 24, 25, 27, 35, 50, 92, 110
DNA damage, 24, 25, 35, 50
drug delivery, 11, 140
drug discovery, 121
drug release, 33, 159, 160
drug resistance, 30
drug targets, 120
drug toxicity, 29

E

E. coli, 11, 12, 13, 15, 30, 31, 33, 42, 56, 57, 58, 59, 60, 62, 63, 64, 65, 66, 67, 68, 69, 70, 71, 72, 73, 76, 78, 91, 92, 94, 96, 97,

99, 106, 109, 110, 112, 113, 114, 116, 134, 141
emulsion, ix, 127, 128, 129, 130, 131, 133, 134, 135, 136, 137, 138, 139, 141, 143, 144, 145, 146, 147, 154, 156, 157, 160
encapsulation, vi, vii, x, 32, 73, 113, 115, 128, 139, 149, 150, 151, 152, 153, 154, 155, 157, 158, 159, 161, 162, 163, 164, 165, 166, 167
energy, ix, 8, 127, 129, 130, 131, 132, 135, 136, 137, 142, 156
environment, viii, x, 11, 31, 66, 89, 93, 113, 124, 128, 149
environmental stress, viii, x, 89, 90, 92, 149
environmental stresses, viii, x, 89, 90, 92, 149
equipment, 67, 90, 92, 94, 97, 98, 99, 112, 138, 156
essential oil, vii, viii, ix, 1, 2, 6, 7, 17, 34, 35, 36, 37, 38, 39, 40, 41, 42, 43, 44, 45, 46, 47, 48, 49, 50, 51, 54, 55, 56, 57, 61, 64, 66, 73, 74, 75, 76, 77, 78, 79, 80, 81, 82, 85, 86, 90, 91, 101, 103, 104, 114, 115, 116, 117, 118, 119, 120, 121, 122, 123, 124, 127, 128, 129, 131, 132, 138, 139, 142, 144, 146, 147, 149, 150, 153, 155, 160, 162, 163, 164, 165, 166
exposure, viii, 11, 13, 18, 20, 53, 55
extraction, 5, 6, 7, 8, 35, 40, 41, 48, 50, 51, 102, 103, 104, 117, 120, 123, 150

F

fat, viii, 53, 64, 65, 66, 69, 70, 71, 72, 73, 77, 78, 83, 87, 98, 99, 101
films, 65, 73, 78, 81, 113, 114, 118
fish, 56, 87, 106, 110, 124
flavor, 5, 74, 75, 78, 101, 130
fluid, ix, 7, 21, 70, 104, 117, 128
fluid extract, 7, 104, 117
food additive, vii, 2, 3, 42, 49, 86, 119, 123
food chain, 111
food industry, viii, 62, 78, 89, 90, 91, 92, 93, 94, 97, 101, 106, 112, 113, 114, 116, 117, 118, 141
food products, x, 85, 142, 149, 151

food safety, 67, 68, 81, 90, 117, 122, 123, 141, 148, 161, 162
food security, 141
food spoilage, viii, 90, 111
foodborne illness, viii, 53, 61, 93
foodborne pathogens, 54, 55, 61, 67, 69, 70, 83, 94, 97, 115, 122, 162
formation, vii, viii, ix, 4, 7, 10, 14, 15, 25, 33, 39, 42, 43, 58, 60, 61, 62, 83, 86, 90, 91, 92, 93, 96, 97, 98, 99, 101, 109, 110, 111, 112, 114, 115, 117, 118, 119, 120, 122, 123, 124, 128, 131, 135, 136, 137, 155, 157
free energy, ix, 128, 136, 142
free radicals, vii, 2, 17, 20
freeze drying, 154, 163
fruits, 73, 94, 96, 97, 108, 141
fungi, 9, 11, 16, 33, 38, 65, 106

G

genes, 12, 24, 27, 29, 34, 45, 50, 56, 57, 61, 62, 91, 101, 110, 111, 117, 124
growth, vii, viii, 10, 13, 15, 16, 24, 33, 42, 54, 55, 61, 62, 63, 65, 66, 69, 73, 76, 77, 79, 94, 99, 100, 106, 109, 115, 117, 137, 141

H

health, vii, 1, 2, 9, 15, 43, 51, 55, 56, 61, 90, 97, 116, 119, 150
hemolytic uremic syndrome, 97
hepatocellular carcinoma, 41, 51
hepatotoxicity, 18, 35, 41, 140
human, vii, 2, 14, 18, 21, 25, 26, 27, 28, 30, 31, 34, 35, 38, 39, 40, 41, 42, 43, 47, 48, 51, 96, 112, 122, 151
hydrogen atoms, 19
hydrogen bonds, 155, 157
hydrogen peroxide, 25
hydrolysis, 7, 31, 32, 59
hydrophobicity, ix, 4, 16, 109, 123, 127
hydroxyl, 4, 9, 16, 17, 19, 30, 31, 109
hygiene, 9, 92, 94, 95, 98, 99, 106

I

immune response, 20
immune system, 20, 21
immunocompromised, 140
immunomodulatory, 38, 131, 141
in vitro, 4, 17, 21, 31, 33, 34, 37, 39, 49, 50, 60, 61, 86, 123
in vivo, 4, 17, 23, 28, 31, 37, 42, 109, 134
including, 6, 10, 14, 33, 54, 59, 61, 67, 82, 90, 103, 141, 150, 151, 155, 157
induction, 23, 25, 33, 35, 36, 60
infection, 12, 14, 15, 20, 45, 50, 141
inflammation, 17, 19, 21, 23, 32, 34, 38, 45, 47, 49
inflammatory mediators, 20, 31, 34, 141
infrared spectroscopy, 140
ingredients, 35, 50, 77, 154, 155, 162
inhibition, 14, 22, 25, 26, 33, 42, 55, 59, 60, 61, 65, 76, 81, 91, 100, 101, 107, 108, 109, 112, 114, 141, 151
injury, iv, 18, 19, 20, 32, 35, 42, 140
interaction effect, 66, 70, 71
interface, 131, 135, 136, 137, 142, 157
interfacial layer, ix, 128, 137
interfacial polymerization, 153, 157
ionic gelation, 156, 164
ions, 28, 58, 59, 70, 136, 156

K

kidney, 12, 18, 19, 48
kinetic model, 150, 159
kinetic modeling, 150, 151
kinetics, 159, 160, 161

L

L. monocytogenes, 30, 33, 56, 60, 62, 63, 66, 67, 68, 73, 74, 78, 92, 94, 95, 97, 100, 106, 109, 112, 113, 114
lactate dehydrogenase, 27, 58
lactic acid, 63, 65, 106, 123
lipid oxidation, 63
lipid peroxidation, 18, 19, 28
liquid phase, ix, 127, 135, 155
liquids, 129, 132, 135, 142, 150

Listeria monocytogenes, viii, 4, 13, 35, 37, 51, 56, 78, 80, 81, 83, 84, 85, 90, 96, 107, 115, 119, 120, 122, 123
liver, viii, 2, 4, 18, 19, 27, 28, 35, 45, 48, 140
liver cancer, 27, 28
liver enzymes, 140
lyophilization, 153, 154

M

materials, 7, 44, 84, 92, 98, 100, 113, 120, 152, 154, 159
matrix, viii, x, 53, 60, 62, 63, 65, 66, 68, 70, 71, 74, 77, 89, 90, 92, 112, 149, 156, 159, 161
meat, vii, viii, 3, 53, 54, 55, 56, 57, 61, 62, 63, 64, 65, 66, 67, 68, 69, 70, 71, 73, 74, 75, 76, 77, 78, 79, 80, 82, 83, 84, 85, 86, 87, 90, 97, 98, 99, 101, 106, 107, 108, 114, 117, 119, 121, 124, 125, 130
meat safety, 54
medicine, 2, 9, 40, 51, 54, 80, 106, 140, 150
melting temperature, 71
membrane permeability, 15, 31, 59, 141
membranes, 16, 17, 57, 76, 141
microbial cells, 90, 92
microbial growth, 94
microbiota, 9, 15, 55, 56, 62, 63, 65
microcapsules, 73, 76, 77, 82, 148, 152, 157, 158, 161, 163, 165
microemulsion, 129
microorganisms, vii, viii, 9, 10, 11, 15, 16, 20, 41, 54, 55, 56, 57, 58, 61, 62, 63, 64, 66, 70, 77, 78, 79, 85, 86, 89, 92, 94, 95, 97, 98, 99, 100, 107, 114, 140, 150
microscopy, 100, 110, 124, 138
microstructure, 133
microwave heating, 48, 122
migration, 21, 31, 41, 50, 65
molecular weight, 3, 135, 155
molecules, viii, ix, 2, 14, 17, 23, 31, 58, 62, 90, 91, 93, 110, 111, 113, 127, 129, 130, 132, 135, 136, 137, 142, 150, 157
morphology, 19, 54, 136, 138, 139, 142, 153

mortality, 11, 13, 24, 71, 98, 140

N
nanocapsules, 64, 113, 152, 164
nanocrystals, 131
nanoparticles, 32, 34, 45, 48, 86, 113, 114, 128, 142, 143, 144, 147, 159, 160, 164, 166
nanostructures, 115
natural antimicrobial, 40, 41, 54, 56, 63, 64, 69, 73, 78, 81, 101, 115, 122
natural compound, 2, 21, 25, 47, 62, 69, 121, 141
natural disinfectant, viii, 90, 106
neurodegenerative diseases, 17, 21, 131
neurodegenerative disorders, 140
nutrients, viii, 64, 89, 90, 93, 95, 97, 99

O
oil, vii, viii, ix, 2, 6, 7, 34, 35, 36, 37, 38, 39, 40, 41, 43, 44, 45, 46, 47, 48, 49, 50, 51, 55, 56, 63, 74, 75, 76, 77, 80, 81, 82, 83, 85, 86, 90, 101, 103, 104, 115, 116, 117, 119, 120, 121, 122, 123, 124, 127, 128, 129, 130, 131, 132, 134, 135, 136, 137, 138, 139, 142, 149, 150, 153, 156, 157, 160
optical microscopy, 138, 139
optical properties, 153
organic compounds, 66
organic solvents, 103
organs, 5, 101, 102
oxidation, ix, 77, 127, 152, 154, 156
oxidative damage, 17, 19
oxidative stress, 4, 17, 18, 19, 28, 29, 30, 34, 35, 38, 42, 44, 45, 47, 48, 49, 77, 140
oxygen, x, 17, 19, 20, 128, 149, 151, 154

P
P. aeruginosa, 4, 11, 13, 14, 29, 32, 33, 42, 45, 50, 80, 85, 91, 100, 107, 109, 111, 112, , 115, 123, 133, 141
partition, 16, 71, 79, 109, 137, 159, 161
pathogens, vii, viii, 2, 9, 10, 11, 12, 13, 14, 16, 36, 53, 54, 55, 56, 57, 58, 59, 60, 61, 62, 63, 64, 65, 66, 67, 69, 70, 71, 74, 75, 76, 78, 83, 84, 90, 94, 95, 97, 113, 115, 119, 122, 124, 141, 150, 151
permeability, 21, 31, 57, 58, 60, 65, 69, 76, 106, 109
pH, viii, 53, 58, 64, 65, 66, 80, 90, 128, 130, 155, 160
pharmaceutical, vii, 1, 5, 28, 30, 33, 35, 80, 128, 129
phenol, vii, viii, ix, 2, 3, 4, 53, 101, 127, 128, 150, 151
physical characteristics, 139, 152
physical properties, 150
physical treatments, 34
physicochemical methods, 153
physicochemical properties, 30, 31, 77, 137, 158, 159, 161
plant-derived compound, 47, 90
plants, vii, viii, ix, 2, 5, 7, 9, 37, 53, 54, 61, 76, 92, 94, 98, 100, 101, 102, 103, 104, 114, 116, 117, 119, 124, 127, 128, 142, 150, 151
polymer, 65, 113, 156, 157, 159, 161
polypropylene, 35, 100, 109, 112, 115
polystyrene, 13, 49, 96, 98, 100, 112
poultry, 56, 69, 76, 79, 83, 86, 107
preparation, iv, vii, ix, x, 32, 56, 68, 94, 128, 133, 150, 151, 156
proliferation, 24, 26, 27, 28, 33, 38, 41, 50, 55
prostate cancer, 4, 27, 40, 41, 50
protection, x, 4, 19, 77, 80, 82, 83, 84, 87, 91, 119, 123, 150, 151, 153, 161
proteins, 17, 21, 25, 26, 27, 28, 40, 55, 59, 60, 64, 65, 70, 91, 92, 95, 97, 99, 110, 118
public health, viii, 90, 92, 141
pulmonary function test, 18, 42
pulmonary tissue damage, 18

R
reactive oxygen, 23, 25, 41, 42, 60, 76, 140
receptor, 26, 33, 91, 110, 117
regression, 68, 70, 71, 72, 159
release kinetics, 162, 163, 165, 166

resistance, viii, 2, 16, 24, 30, 50, 57, 59, 64, 68, 69, 70, 71, 74, 79, 89, 95, 100, 109, 111, 115, 119, 120, 125, 131
respiratory syncytial virus, 14
response, 20, 23, 27, 35, 37, 39, 48, 70, 77, 81, 93, 116
rheology, 130, 137, 139, 142, 153

S

safety, viii, 4, 34, 51, 53, 54, 56, 66, 67, 68, 69, 75, 78, 80, 86, 112, 117, 141, 151, 161
Salmonella, viii, 4, 11, 13, 25, 33, 38, 40, 43, 49, 56, 62, 63, 64, 67, 68, 69, 70, 74, 80, 82, 83, 84, 85, 87, 90, 92, 94, 96, 97, 98, 106, 107, 109, 113, 115, 116, 118, 121, 123, 124, 133
sanitary, 95
scanning electron microscopy, 138, 139, 158
scarcity, 154
science, 38, 40, 79, 82, 83, 85, 117, 118
sensing, 4, 33, 36, 50, 62, 91, 115, 118, 123, 125
shelf life, 56, 67, 75, 76, 85, 87, 151
showing, 16, 25, 106, 109, 110
sodium, 23, 25, 34, 56, 65, 110, 113, 122
sodium dodecyl sulfate, 113, 122
solubility, ix, 5, 28, 30, 31, 32, 33, 41, 63, 64, 66, 113, 127, 129, 135, 137, 142, 151, 153, 157, 159, 161
species, 6, 11, 17, 23, 25, 36, 41, 42, 49, 60, 61, 69, 76, 83, 96, 98, 102, 110, 115, 118, 119, 124, 140, 150
spray drying, 78, 153, 154, 155, 161, 163, 164, 165
stabilizers, vii, ix, 128, 130
Staphylococcus aureus, 4, 10, 11, 31, 32, 33, 38, 55, 56, 58, 59, 66, 68, 80, 81, 83, 92, 95, 97, 106, 107, 109, 110, 112, 113, 114, 118, 122, 124, 141
steel, 13, 38, 49, 92, 95, 96, 97, 99, 100, 110, 112, 113, 115, 119, 120, 122, 123, 141

storage, 57, 62, 64, 75, 86, 87, 92, 112, 130, 131, 140, 141, 152, 161
structure, ix, 3, 4, 19, 36, 60, 65, 69, 70, 91, 102, 110, 127, 132, 155
surface area, 132
surface properties, 100, 124
surface tension, 132, 135
surfactant, 129, 130, 131, 132, 134, 135, 137, 142, 156
susceptibility, ix, 51, 81, 112, 127, 130
synergistic effect, 29, 30, 58, 65, 74, 76
synthesis, 2, 21, 30, 34, 37, 38, 56, 58, 62, 91, 109, 110, 111, 112, 114, 117

T

techniques, 5, 7, 8, 35, 42, 78, 102, 103, 129, 133, 138, 140, 142, 156, 158, 161
temperature, viii, x, 5, 54, 64, 65, 67, 68, 69, 70, 71, 72, 73, 82, 85, 90, 97, 99, 100, 119, 120, 130, 132, 136, 149, 151, 154, 155, 156, 157
therapy, 2, 9, 19, 23, 24, 29, 30, 33, 58, 150
thermal destruction, 82
thermal resistance, 57, 69, 70, 71, 72, 74, 78
thermal stability, 161
thermal treatment, 65
thermodynamic equilibrium, ix, 128, 159
toxin, 49, 56, 86, 99, 106, 116, 119, 123
transmission, 70, 71, 94, 125, 138, 139, 158
transmission electron microscopy, 138, 158
treatment, 4, 11, 18, 19, 21, 22, 23, 32, 45, 46, 56, 57, 58, 59, 60, 62, 63, 67, 68, 74, 76, 78, 97, 110, 141

U

ultrasound, 8, 40, 103, 129, 134
urinary tract, 9, 15, 38, 46
urinary tract infection, 15, 38, 46
UV light, 56
UV radiation, 17

V

vegetables, 55, 67, 80, 94, 96, 97, 108, 122
virus infection, 51
viruses, 11, 14, 16, 33, 47, 76, 141

volatility, ix, 30, 63, 64, 112, 113, 114, 127, 128, 142, 151

W

water, viii, ix, 3, 7, 11, 28, 30, 31, 53, 63, 64, 67, 70, 71, 77, 79, 86, 94, 95, 98, 99, 127, 128, 129, 130, 132, 135, 136, 137, 142, 151, 153, 154, 156, 157, 160

wettability, 97, 135, 137

worldwide, 2, 9, 13, 14, 15, 23, 24, 38, 90, 94, 140

wound infection, 29, 32, 45

Y

yield, 8, 40, 48, 49, 76, 103, 123, 161